Natural Computing Series

Founding Editor

Grzegorz Rozenberg

Scope

Natural Computing is one of the most exciting developments in computer science, and there is a growing consensus that it will become a major field in this century. This series includes monographs, textbooks, and state-of-the-art collections covering the whole spectrum of Natural Computing and ranging from theory to applications.

More information about this series at https://link.springer.com/bookseries/4190

Neil Urquhart

Nature Inspired Optimisation for Delivery Problems

From Theory to the Real World

 Springer

Neil Urquhart
School of Computing
Edinburgh Napier University
Edinburgh, UK

ISSN 1619-7127
Natural Computing Series
ISBN 978-3-030-98110-5   ISBN 978-3-030-98108-2 (eBook)
https://doi.org/10.1007/978-3-030-98108-2

This Springer imprint is published by the registered company Springer Nature Switzerland AG
The registered company address is: Gewerbestrasse 11, 6330 Cham, Switzerland

Dedicated to Jamie, who might be interested in this when he grows up.

Preface

The aim of this book is to attempt, in a small way, to bridge the gap between Nature Inspired Computing research and real-world applications. The domain that we utilise is that of delivery problems, that being an area that has increasing relevance to many of us in our day-to-day lives. I first encountered Nature Inspired Computing in the form of Evolutionary Algorithms as an undergraduate student. In my class, an assessment had been set that involved writing a program to solve the Travelling Salesperson Problem, and the suggested technique involved concurrent programming, recursion and trees. I'm not sure who introduced me to the EA, but I soon realised that I could code an EA to find an answer to the TSP far more quickly than my classmates could build their assessment. Not only did I get the assignment completed far quicker than anyone else, but I also avoided having to deal with such issues as recursion, trees and parallelism. My lecturer (who later became a valued colleague) was not overly impressed. This initiation into Nature Inspired Computing captured my imagination and started me on a journey that lead to an Undergraduate dissertation, a Ph.D. thesis and then many happy years lecturing at Edinburgh Napier University.

By the end of this book, I hope that the reader will have learned not only about NIS, but also grasped how to apply it to solve useful problems. I also hope that some of my enthusiasm for Nature Inspired Computing will have been passed on.

I have attempted to take a practical approach within this book, making sure that emphasis is placed on implementation issues as much as on theory.

Edinburgh, UK
September 2021

Neil Urquhart

Acknowledgements

This book would not have been possible without the support and encouragement of past and present colleagues at Edinburgh Napier University and elsewhere who have encouraged, corrected and tolerated me over the years.

Thanks to

- Leanne Clyde
- Thomas Farrenkopf
- Achile Fonzone
- Michael Guckert
- Emma Hart
- Silke Höhl
- William Hutcheson
- Rob Kemmer
- Jon Kerridge
- Ben Paechter
- Simon Powers
- Bob Rankin
- Bryden Ritchie
- Peter Ross
- Kevin Sim
- Katy Thierens
- Jennifer Willies

A special thanks must go to David Bird who inspired me during my first steps into Computing Science many years ago. I must also thank the many students at Edinburgh Napier University who have inspired me over the years.

Much of this work was originally published and presented at the annual *EvoStar* conference. I must thank the organisers, delegates and reviewers for creating an environment that has been educational and inspirational over the last 20 years.

And finally, a massive thank you to my wife, Siân, for her tolerance, understanding and encouragement over the years.

A Note on the Software Provided with This Book

The Java software examples presented in this book are available from a repository located at https://github.com/NeilUrquhart/book. Readers are invited to download and run the examples and also to modify and experiment with them as they see fit.

Also provided within the repository are several sets of slides and some tutorial notes for use by anyone wishing to teach a course based on the topics contained in this book.

Please see the read-me file within the repository for full details.

Contents

Acronyms

AGI	Artificial General Intelligence
AI	Artificial Intelligence
API	Application Program Interfaces
B2C	Business to Consumer
CSV	Comma Separated Values
CVRP	Capacitated Vehicle Routing Problem
CVRPTW	Capacitated Vehicle Routing Problem with Time Windows
DVRP	Dynamic Vehicle Routing Problem
EA	Evolutionary Algorithm
EC	Evolutionary Computation
EV	Electric Vehicle
GIS	Geographical Information System
GPS	Global Positioning System
GPX	GPS exchange
ILP	Integer Linear Programming
KML	Keyhole Markup Language
MAP-Elites	Map Archive of Phenotypic Elites
MD	Micro-Depot
ML	Machine Learning
MS	Milliseconds
NN	Nearest Neighbour
NP	Non-Polynomial
ONS	Office of National Statistics
OR	Operations Research
OSM	OpenStreetMap
SBR	Street Based Routing
SVRP	Stochastic Vehicle Routing Problem
TSP	Travelling Salesperson Problem
VRP	Vehicle Routing Problem
VRPSTW	Vehicle Routing Problem with Soft Time Windows
VRPTW	Vehicle Routing Problem with Time Windows

Part I
Interesting Problems and How to Solve Them

This section examines fundamental problems such as the Travelling Salesperson Problem and related problems such as the Vehicle Routing Problem and Vehicle Routing Problem with Time Windows. We explore why it is that such problems are easy to understand, yet computationally difficult to solve. We introduce more traditional algorithmic techniques such as the 2-opt heuristic and the Clarke–Wright algorithm prior to introducing evolutionary inspired techniques such as the Evolutionary Algorithm and MAP-Elites.

This section follows a chronological order, commencing with the Travelling Salesperson Problem and progressing through problems and case studies to cover vehicle routing problems.

Chapter 1
The Travelling Salesperson Problem

Abstract This chapter introduces the Travelling Salesperson Problem (TSP) which underpins almost all other delivery type problems. Within the TSP, a route must be found that visits a set of locations within the shortest possible distance, and each location must be visited once. The TSP has the useful properties of being very easy to understand, whilst presenting a considerable computational challenge to solve as it does not scale up well. If we are to understand more complex problems, then a review of the TSP is a useful starting point. We examine a number of heuristics (such as Nearest Neighbour and 2-opt) which may be used to generate solutions. Through a case study, we discuss the software engineering practicalities of implementing TSP solvers and compare the performance of the heuristics discussed.

1.1 A Brief History of the Travelling Salesperson Problem

The Travelling Salesperson Problem[1] (TSP) has a long history (Cook 2012; Cummings 2000), and the formulation of the modern TSP is widely attributed to Karl Menger who introduced the Das Botenproblem at a mathematics conference in 1932 (Menger 1932). Mengers' problem involved finding the shortest route for postal messengers who must visit a set of points. The first known usage of the name Travelling Salesperson Problem was by Julia Robinson in 1949. Robinson had difficulties in solving the TSP and it was not until 1954 that researchers Dantzig et al. (1954) adopted an Operations Research (OR)-based approach and successfully solved a range of instances (including that proposed by Julia Robinson in 1949), using linear programming. The importance of the TSP as a starting point for this work cannot be underestimated, as well as being a significant theoretical problem instances have incorporated real-world problems including the delivery of fuel oil (Dantzig et al. 1954; Dantzig and Ramser 1959). As computing hardware increased in

---

[1] The TSP was originally described as the Travelling Salesman, but many recent publications refer to it as the Travelling Salesperson Problem. For consistency, this book uses the term Salesperson, but many of the cited publications use Salesman.

© Springer Nature Switzerland AG 2022
N. Urquhart, *Nature Inspired Optimisation for Delivery Problems*,
Natural Computing Series, https://doi.org/10.1007/978-3-030-98108-2_1

power and availability, its use for solving problems such as the TSP became obvious. In 1958, Bocks (1958) described solving the TSP using an algorithm implemented on an IBM 650 computer .

1.2 Problem Description

The TSP may be described as follows:

> A Salesperson must visit at a number of cities, each city must be visited once and only once, with the Salesperson ending up back at the start point. The aim being to find the shortest route that will enable the Salesperson to make their visits.

The problem of finding the optimum order of set of visits remains at the core of most delivery problems.

A valid answer to the TSP takes the form of *permutation* of the cities, a permutation being a set of objects, where each object occurs once and once only in the set, nothing is duplicated and nothing is missed out. Thus, a permutation represents a *tour* that takes in each city once. Such tours are more formally known as *Hamiltonian Circuits* after the nineteenth-century Irish mathematician W. R. Hamilton.

For a 5-city TSP with cities A–E, examples of valid answers would be ABCDE or BEACD but not ABCD or BCDEAC. Figure 1.1 shows a simple example of a 5-city TSP tour.

Solving the TSP should be reasonably straightforward; evaluate each possible permutation of cities (by measuring the distance of each route) and note the permutation that results in the shortest distance. The number of possible solutions to a TSP problem is the number of valid permutations of cities. The number of

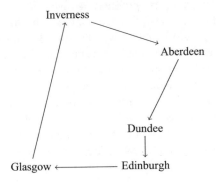

Fig. 1.1 Here, we see a very simple 5-city TSP. The solution takes the form of a tour that visits each city once. The quality of the solution is based upon the distance that must be travelled in order to visit each city in the order specified. The reader should remember that there are 5! (120) possible ways of touring these cities of which this is only one

permutations that can be constructed for a set of n cities is $n!$ (n factorial) which is simply $n * n - 1 * n - 2 \ldots n - n$. In plain language, if we have 5 cities, the number of permutations is $5 * 4 * 3 * 2 * 1$ which is 120. So if our Salesperson has 5 cities to visit, we need to measure then the length of 120 routes and pick the shortest. Measuring the length of 120 routes around 5 cities is a trivial task, and computationally, we could write a simple program to solve the 5-city TSP. The set of all possible solutions to a problem is known as the *solution space*.

What happens if we add a sixth city? Rather alarmingly the number of possible solutions increases by 600% to 720, and here lies the crux of the problem... *scalability*. Consider the following table:

Cities	Possible solutions
5	120
10	3628800
15	1.30767×10^{12}
20	2.4329×10^{18}
25	1.55112×10^{25}
30	2.65253×10^{32}
35	1.03331×10^{40}
40	8.15915×10^{47}
45	1.19622×10^{56}
50	3.04141×10^{64}
60	8.32099×10^{81}

As soon as the number of cities rises beyond 10, we're into very large numbers of solutions to search. In fact, the number of atoms in the visible universe is estimated to be in the range of 10^{78}–10^{82} (Villanueva 2018), and it follows that if we have a TSP with 60 or more cities, then there are more solutions than atoms in the universe. As we will see, this increase in the number of possible solutions poses a significant problem for us in attempting to solve TSP instances. The TSP is often described by researchers as being *NP-Complete*. Problems that are considered NP-Complete are those which, although they are possible to solve, cannot be solved using an exhaustive search in a reasonable time. This is normally defined as the problem being solvable in *polynomial time*, that is to say there is a polynomial link between the problem size and the time to solve.

If a problem can be solved in polynomial time, then it scales up easily, suppose we had a delivery problem where the number of operations to find the optimum solution was the number of deliveries multiplied by 2 (we can assume that each operation takes the same amount of time to execute). So 10 deliveries would be solved in 20 operations, 20 in 40 operations and so on. It would be possible to increase the problem size to very large instances and still solve within a reasonable time, in particular adding a single delivery to a problem instance would always increase the time to solve by 2 operations. In an NP-complete problem, the time to solve increases dramatically as the problem becomes larger. Adding an extra city to a small TSP instance results in an increased time to solve, adding the city to a larger instance results in a much larger increase in solving time.

If a problem such as TSP is NP-Complete, then any problems which incorporate it are themselves NP-Complete. In our domain of delivery problems, most problems require the solving of a TSP as part of their solution; therefore, most of the problems that we are examining in this book are NP-Complete.

But what does this mean in practical terms when we are designing and implementing solvers?

- We cannot adopt an *exhaustive search* approach for anything but the smallest problem instances.
- We may never find the optimum solution.
- We must encourage the end user to be satisfied with a high-quality solution, rather than the optimal solution.

In the case of many real-world problems, the difference between the optimum solution and a high-quality solution may be little in practical terms. For example, if the optimum solution for a TSP found a route around a city which is 2 min shorter than an easily found high-quality solution, could be argued that the saving of two minutes is insignificant in practical terms (unexpected traffic congestion could easily delay the Salesperson by more than two minutes).

It is against this background of difficult NP-Complete problems that will attempt to make use of nature inspired techniques, not to find optimal solutions to problems, but to find the best solution that we can solve in a reasonable time, using reasonable[2] hardware.

1.3   Techniques for Solving the TSP

A search through TSP published literature reveals many different techniques for finding a solution to the TSP (the reader is directed to Cook 2012 for an overview), and we will investigate a exhaustive search, the Nearest Neighbour heuristic (Knuth 1973) and the 2-opt iterative heuristic (Flood 1956; Croes 1958).

Exhaustive Search

The exhaustive search approach (sometimes known as *brute force* search) creates every single possible solution and evaluates each of them. As the heuristic evaluates each solution, it keeps a note of the best solution found so far. In the case of the TSP, the number of possible solutions is the $n!$ (where n is the number of cities).

---

[2] I very deliberately use the ambiguous term *reasonable*. What is a reasonable time to a patient user is an eternity to an impatient user. Hardware specifications constantly improve and a reasonable computer of this year may be regarded as underpowered and obsolete by next year.

The pseudocode in Algorithm 1 shows the outline of the exhaustive search. We track the shortest route found so far using $bestSoFar$ (line 1)—note this simple version only tracks the length of the shortest route and a more complex implementation would also keep a record of the solution itself. The function $getNextSolution()$ returns the next valid permutation, repeated calls to $getNextSolution()$ return each possible permutation until they have all been returned and then $null$ (line 3) is returned.

This approach has the advantage that because $every$ possible solution is evaluated, it is guaranteed to find the shortest route. But this advantage is countered with the disadvantage that this approach has scaling issues, as we shall see it is impractical for all but the smallest problem instances.

Algorithm 1 TSP—Exhaustive Search

1: $bestSoFar = \infty$
2: $solution = getNextSolution()$
3: **while** solution != null **do**
4:     **if** $tourLength(solution) < bestSoFar$ **then**
5:         $bestSoFar = tourLength(solution)$
6:     $solution = getNextSolution()$
7: **return** $bestSoFar$

The Nearest Neighbour Heuristic

The nearest neighbour (NN) heuristic is a *approximation algorithm* that constructs a single solution to the problem. This approach was first discussed in Knuth (1973), where it was referred to as the Post Office problem. This solution whilst *probably* a reasonable solution is not guaranteed to be the optimal solution.

Algorithm 2 describes the basic structure of NN. Initially, all of the cities are not visited (line 1, Fig. 1.2a), and the first city in the solution is selected randomly (line 3, Fig. 1.2b); note that the $removeRand()$ function removes a random city from the list of cities not yet visited. Whilst there are cities still to be visited (line 5), the algorithm checks the distance between the current city (the latest added to the solution) and each remaining city yet to be visited (line 8, Fig. 1.2c and d). The nearest unvisited city is then added to the tour (line 11), and this continues until all of the cities have been added to the solution.

The NN heuristic typically returns a sub-optimal solution (see Fig. 1.2), and a frequent problem is that a city is left out of the tour earlier and ends up being added by default at the end as the last visited city. In our example (Fig. 1.2f), city A ends up at the end of the tour as it is too far away from cities B and C to be incorporated earlier. One method of assessing the effectiveness of an NN solution is to compare the lengths of the final arc (in this case $A - E$) with the average length, and if the final

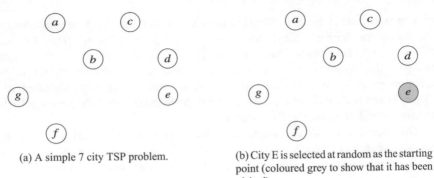

(a) A simple 7 city TSP problem.

(b) City E is selected at random as the starting point (coloured grey to show that it has been visited).

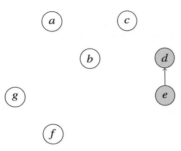

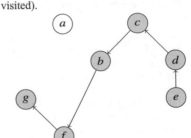

(c) D is closest to E, so it is added next.

(d) After E, the closest is C then B, F and G.

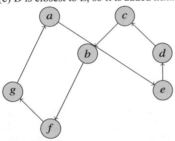

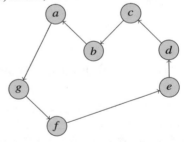

(e) The last unvisited city is A, which is added at the end of tour

(f) The tour created by NN (figure 1.2e is obviously sub-optimal, a simple change such as this can improve upon the NN tour.

Fig. 1.2 The Nearest Neighbour heuristic in action. Figure 1.2f shows that NN solution could be improved

arc is significantly longer, then that suggests that the solution is sub-optimal due to the final city being misplaced within the solution. As an example, Fig. 1.2f shows a solution to the problem that does not display this characteristic.

Some variants of the TSP specify a starting city, which removes the need to choose a starting point at random. Where the starting city is chosen at random, this ensures that the algorithm is *stochastic*. When a *stochastic* algorithm is employed, the user takes a risk that the algorithm will produce a sub-standard answer on a specific run.

This may be overcome to some extent by executing the algorithm several times and accepting the best result.

Algorithm 2 TSP—Nearest Neighbour

1: *notVisited* = 1 . . . *n*
2: *solution* = ∅
3: *current* = *removeRand*(*notVisited*)
4: *solution* ∪ *current*
5: **while** *notVisited*! = ∅ **do**
6:    *best* = ∞
7:    **for** *possible* : *notVisited* **do**
8:       **if** *dist*(*current*, *possible*) < *best* **then**
9:          *next* = *possible*
10:    *notVisited* = *notVisited* − *next*
11:    *solution* ∪ *next*
12:    *current* = *next*
13: **return** *solution*

The 2-opt Heuristic

The 2-opt heuristic (Croes 1958) is a *local search* algorithm that iteratively improves a TSP solution. Starting from an initial random solution, 2-opt makes changes to the solution, and those changes which result in an improved (shorter) solution are adopted, whilst those that make a tour longer are rejected.

Consider the tour [*ABEDCFG*] as follows:

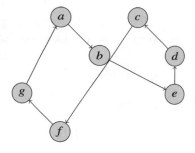

We can see that this tour is clearly sub-optimal, but by reversing the order of cities *EDC* to *CDE*, we see the following improved solution:

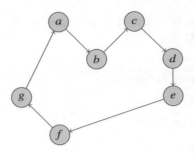

The 2-opt heuristic operates by selecting pairs of cities and reversing the order of the tour between them. In some cases (such as that shown above), this results in an improvement to the tour. The basic principle of iterative improvement is shown in Algorithm 3. Firstly, a random solution is generated (line 1), and a new solution (*solution'*) is created by making a random change to *solution* (line 3). If the new solution represents an improvement (line 4), then the new solution is adopted (line 5). We continue to attempt to improve the solutions until some stopping criterion is met (line 2). We will return to this basic design pattern at several points within this book.

Algorithm 3 Iterative improvement

1: *solution* = *randomSolution*()
2: **while** !*done* **do**
3:     *solution'* = *randomChange*(*solution*)
4:     **if** *quality*(*solution'*) < *quality*(*solution*) **then**
5:        *solution* = *solution'*
6: **return** *solution*

In the case of 2-opt, we modify the solution by making the 2-opt swap shown above and described in listing Algorithm 4. Let's assume that we have a 10-city tour as follows:

$$ABCDEFGHIJ$$

Now we select positions 4 and 8 for the swap:

$$ABC\textbf{DEFGH}IJ$$

Applying the swap, the tour would become

$$ABCHGFEDIJ$$

There is no guarantee that a particular swap will improve a tour, and the power of 2-opt is that it attempts many swaps and retains those that improve the tour and discards those that are not an improvement.

The complete 2-opt algorithm is shown in listing Algorithm 5. The two for loops (lines 5 and 6) select the sections to swap as follows:

i	j	Swap
0	1	**AB** CDE
0	2	**ABC** DE
0	3	**ABCD** E
...	...	...
1	2	A **BC** DE
1	3	A **BCD** E
...	...	...

Each selection of cities is swapped (line 7), and if the resulting tour is shorter, then the modification is retained (lines 8 and 9). Progressing through each pair of cities, as described above, may lead to an improvement, but further improvement may be possible by repeating the search. We repeat the search (line 3) until none of the swaps that are attempted result in an improvement to the solution.

Algorithm 4 2-opt swap

1: **Procedure** $swap(tour, i, j)$
2: **for** $x = 0; x < i; x + +$ **do**
3:    $result[x] = tour[x]$
4: **for** $x = jx >= i; x - -$ **do**
5:    $result[x] = tour[x]$
6: **for** $x = jx < solution.length; x + +$ **do**
7:    $result[x] = tour[x]$
8: **Return** $result$
9: **EndProcedure**

Algorithm 5 2-opt

1: $solution = randomSolution()$
2: $finished = false$
3: **do**
4:    $improved = false$
5:    **for** $x = 0; x < solution.length; x + +$ **do**
6:       **for** $y = x + 1y < solution.length; y + +$ **do**
7:          $solution' = swap(solution, x, y)$
8:          **if** $dist(solution') < dist(solution)$ **then**
9:             $solution = solution'$
10:            $improved = true;$
11: **while** $improved$
12: **Return** $solution$

Evolutionary-Based Methods

This section would not be complete without discussing the application of *Evolutionary Algorithms* EAs to the TSP (Brady 1985; Johnson 1990; Freisleben and Merz 1996). We will fully discuss the use of EAs in the context of Vehicle Routing Problems (VRPs), which are the main focus of this book, in Chap. 3. For completeness, a brief description of an EA is given here.

EAs utilise a *population*-based approach, and they maintain a *population* of multiple solutions rather than building a single solution. Each member of the population has a *fitness* value that represents *quality* of the solution. In the case of a TSP, the fitness is normally the distance of the TSP, and those solutions with a lower distance (i.e. the better solutions) have a lower fitness value. This allows us to identify which members of the population represent better solutions for our TSP.

New solutions are created by selecting pairs of solutions (with a bias towards higher quality solutions) and creating a new *child* solution by combining elements of both parents. This operation is known as *crossover*. One of the challenges of applying an EA to a TSP is ensuring that when a child is created by crossover, it still contains a valid solution to the TSP (i.e. one that visits each city exactly once) (Watson et al. 1998; Varadarajan et al. 2020; Varadarajan and Whitley 2021; Mukhopadhyay et al. 2021).

Having created a new solution using crossover, a small random change, known as a *mutation*, is applied to the child. The child is copied back into the population replacing a solution of lower quality. This process is known as a *generational cycle* and is repeated many thousands of times, allowing the members of the population to improve. The EA has the advantage that a diverse population can be searching in many areas of the solution space simultaneously.

1.4   Software Engineering Considerations

It is at this point that we need to consider the software engineering aspects of solving problems such as the TSP. It is good practice to implement our solver in a manner that decouples the solver from the rest of the system. This allows for easier substitution of different heuristics, in order to assess their suitability for a given problem instance. Object-Oriented languages are particularly appropriate for such implementations such as that proposed in Fig. 1.3. The *Problem* class contains details of the cities and the means of calculating the distances between them (in this simplified example, we omit the methods required for loading the problem data). The *TSP Heuristic* class has one principle method, *solve()* which finds an answer to the Problem linked to the solver. Our *Application* class contains all of the code required to create the *Problem* and *TSP Heuristic* objects and deals with any user interaction required.

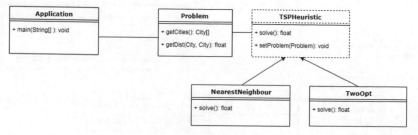

Fig. 1.3 An example of separating the heuristic from the problem and demonstrating the manner in which multiple heuristics may be incorporated

By making *TSPHeuristic* an abstract class, we may then create any number of concrete classes that solve the problem in different ways. This implementation is especially useful for problems such as the TSP where many heuristics are available and it may be desirable to use different heuristics in specific situations.

Within this book, we will attempt to keep heuristics separate from problem data and methods in order to allow for easy experimentation with differing heuristics. In a larger software project, keeping specialist heuristics separate from the rest of the software allows easier development and maintenance of the heuristics by specialists and makes it easier to hide their complexity from other members of the development team.

1.5 A TSP Case Study

This case study encompasses a problem that has puzzled young children across the world, how does Santa Claus choose a route to undertake Christmas present deliveries? Whilst this might seem like a trivial example, this problem has a number of properties which make it useful for investigating at this stage.

Santa Claus delivers presents from his grotto using his sleigh. He leaves the grotto laden with presents and visits the house of each child, delivering presents, before returning to the grotto. This problem may be considered an instance of the Travelling Salesperson Problem. Currently, Santa uses the nearest neighbour technique to work out his delivery plan.

Aware that some optimisation of his route would be useful, Santa has decreed that two optimisation algorithms are to be evaluated as follows:

- Exhaustive Search (Listing 1.1).
- 2-opt (Listing 1.3).

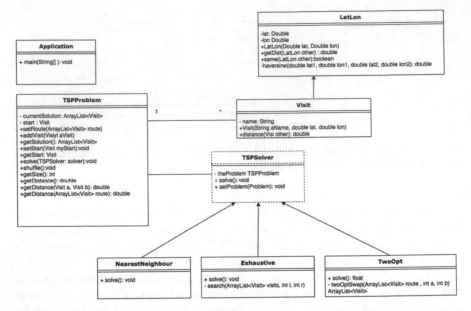

Fig. 1.4 The Santa Claus TSP solver as implemented (see the accompanying code for a full listing)

For comparison purposes, Nearest Neighbour (Listing 1.2) will be used as a baseline to see how the current routes are improved upon.

Santa's Elves have implemented all three algorithms, and a class diagram of their implementation may be found in Fig. 1.4. As discussed in Sect. 1.4, the implementation of the algorithms and the problems has been kept separate. A complete Java listing may be found in the repository that accompanies this book.

The implementation for exhaustive search (Listing 1.1) uses a recursive method (*search*() line8) to create an evaluate every possible permutation of visits.

```
1   public void solve(){
2       lowestDist = Integer.MAX_VALUE;
3       search(theProblem.getSolution(), 0, theProblem.getSize()
        -1);//Start generating permutations
4       theProblem.setRoute(bestRoute);
5   }
6
7   //Recursive search function
8   private void search(ArrayList<Visit> visits, int l,int r)
9   {
10      if (l == r){//completed - nothing left to search
11          theProblem.setRoute(visits);
12          double len = theProblem.getDistance();
13          if (len < lowestDist){//If current is lowest so far
14              lowestDist = len;
15              bestRoute = (ArrayList<Visit>)visits.clone();
16          }
17      }
18      else
19      {
20          for (int i = l; i <= r; i++)
21          {
22              Collections.swap(visits, l, i);//Swap
23              search(visits, l+1, r);//Recursion
24              Collections.swap(visits, l, i);//Undo swap
25          }
26      }
27  }
```

Listing 1.1   The exhaustive search algorithm

```
1   public void solve() {//Solve the problem
2       ArrayList<Visit> newRoute = new ArrayList<Visit>();
3       ArrayList<Visit> notVisited = (ArrayList<Visit>)
        theProblem.getSolution().clone();//Visits to be added
4       Visit current = theProblem.getStart(); //Start
5       while(notVisited.size()>0){//While visits still to add
6           Visit next=null;
7           double bestD= Double.MAX_VALUE;
8           for(Visit possible: notVisited){//Check unvisited
9               if (theProblem.getDistance(current, possible)<
            bestD){
10                  next = possible;
11                  bestD = theProblem.getDistance(current, next
                );
12              }
13          }
14          notVisited.remove(next);//Add the closest visit
15          newRoute.add(next);
16          current = next;
17      }
18      theProblem.setRoute(newRoute);
19  }
```

Listing 1.2   The Nearest Neighbour algorithm

```
1    public void solve(){
2        boolean improved = true;
3        //repeat until no improvement is made
4        while(improved == true){
5            improved= false;
6            double best_distance = theProblem.getDistance();//
  The starting solution
7
8            for(int i=0; i< theProblem.getSize(); i++){
9                for (int k = i + 1; k < theProblem.getSize(); k
  ++) {//loop through each pair of visits
10                   ArrayList<Visit> new_route = twoOptSwap(
  theProblem.getSolution(),i,k);//Undertake the swap
11                   double new_distance = theProblem.getDistance
  (new_route);
12                   if (new_distance < best_distance) {//If the
  modified route is shorter (ie the swap has worked)
13                       theProblem.setRoute(new_route);//The
  modified route now becomes the current route
14                       improved = true;
15                       best_distance = new_distance;
16                       break;//Break out of the loops
17                   }
18               }
19               if (improved)
20                   break;//Break out of the loops
21           }
22       }
23   }
24
25   private ArrayList<Visit>  twoOptSwap(ArrayList<Visit> route,
  int a, int b) {
26       //Perform a 2-opt swap based on the visits positions a
  and b
27       //Return the modified route in a new arrayLisy
28
29       ArrayList<Visit> modifiedRoute = new ArrayList<Visit>();
30       Visit[] old = new Visit[route.size()];
31       old = route.toArray(old);
32       for(int c=0; c <= a; c++)
33           modifiedRoute.add(old[c]);
34
35       for(int c= b; c > a; c--)
36           modifiedRoute.add(old[c]);
37
38       for(int c= b+1; c < route.size(); c++)
39           modifiedRoute.add(old[c]);
40
41       return modifiedRoute;
42   }
```

Listing 1.3 The 2-opt algorithm

The Elves test the algorithms on a sample dataset, based on 50 visits across the world, starting and ending at the Grotto in Lapland (see Fig. 1.5). The Elves execute each algorithm 10 times on each algorithm (see the *CompareApp.java* listing in the supplied code) and measure the average execution time and distance over 10 runs.

When measuring execution time on an MS Windows, Linux or MacOS-based computer, our application shares the CPU and other resources with many other

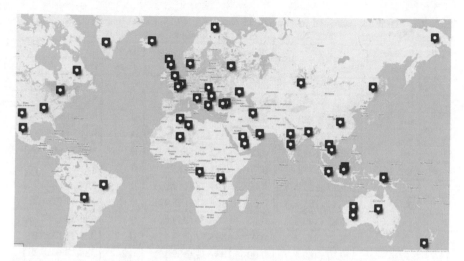

Fig. 1.5 The 50 visits used to generate the test data. *Base map and data from OpenStreetMap and OpenStreetMap Foundation* (https://www.openstreetmap.org/copyright)

processes being run by the operating system. Our execution time may, therefore, be affected by whatever else is being executed concurrently. Averaging over multiple runs helps minimise the effects of other processes, but the Elves should exercise caution where times are very similar.

The results for the first set of tests carried out by the Elves may be seen in Table 1.1. The first thing that the Elves notice is that the time taken for the Exhaustive search dramatically increases as the number of visits increases. For 12 visits, the average time to solve was just over half a minute; for 13 visits, the average time was approximately 9 min; for 14 visits, they had run out of patience awaiting the results and the exhaustive search algorithm was not used for the remaining problems. The other solvers all produced answers within 1 ms.

In terms of performance, the Elves noted the best result obtained over 10 runs and the average. For the Exhaustive and Nearest Neighbour heuristics, the results are always the same, but as 2-opt commences from a random solution and improves it, then the results will sometimes vary (hence the difference between the best and the average). Not surprisingly, the Exhaustive algorithm gives the best overall result, but the Elves have to discount it as it clearly won't scale up as the number of visits increases. There is little practical difference in times between Nearest Neighbour and 2-opt. Examples of solutions generated with the Nearest Neighbour and 2-opt algorithms may be seen in Fig. 1.6a and b.

The Elves continue their experiments using Nearest Neighbour and 2-opt, and their results are shown in Table 1.2. As the number of visits increases, two things become apparent; firstly, 2-opt can find better solutions than Nearest Neighbour, and secondly, 2-opt begins to take longer to run as the number of visits is increased (in

Table 1.1 The results obtained for the first 13 test instances. Note that times (milliseconds) are averaged over 10 runs and distances are presented as <min(average)> over 10 runs

Visits	Exhaustive		NearestN		TwoOpt	
	Time	Dist	Time	Dist	Time	Dist
1	0	64.11 (64.11)	0.00	64.11 (64.11)	0	64.11 (64.11)
2	0	68.93 (68.93)	0.00	68.93 (68.93)	0	68.93 (68.93)
3	0.1	69.02 (69.02)	0.00	69.02 (69.02)	0	69.02 (69.14)
4	0	93.23 (93.23)	0.00	93.23 (93.23)	0	93.23 (93.87)
5	0.1	108.77 (108.77)	0.00	112.35 (112.35)	0.1	108.77 (111.04)
6	0.2	125.68 (125.68)	0.00	151.8 (151.8)	0.1	125.68 (148.53)
7	0.9	169.27 (169.27)	0.00	169.27 (169.27)	0	169.27 (193.92)
8	3.4	195.47 (195.47)	0.00	195.47 (195.47)	0	195.47 (217.22)
9	24.9	249.85 (249.85)	0.00	254.66 (254.66)	0	269.58 (288.79)
10	260.5	328.25 (328.25)	0.00	328.77 (328.77)	0	328.77 (356.88)
11	3007.3	389.17 (389.17)	0.00	389.17 (389.17)	0.2	389.17 (418.87)
12	38656.9	389.23 (389.23)	0.00	484.15 (484.15)	0.1	405.76 (438.6)
13	519832.6	514.51 (514.51)	0.00	661.8 (661.8)	0.1	514.51 (556.84)

addition to that, 2-opt needs to be run multiple times in order to maximise the chances of finding a good solution).

It occurs to one of the Elves that Nearest Neighbour and 2-opt could be combined into a hybrid problem solver. Rather than 2-opt commencing with a random solution, it is initialised using the Nearest Neighbour solution. The Nearest Neighbour solution is unlikely to be optimal, but it is a far better solution than the random solution. The role of 2-opt is no longer to construct a solution from scratch, but to improve the Nearest Neighbour solution. Because the Elves adopted good-quality software engineering strategies, the creation of the new solver is very simple (see Listing 1.4). The results obtained with the hybrid solver may be seen in Table 1.3. The Elves notice two things about the Hybrid solver;

- That in all but two cases it finds the best result of the three solvers.
- That it solves the problem far more quickly than the 2-opt solver and because there is no random initialisation, it is not necessary to undertake multiple runs.

Figure 1.6c shows the improvements made by the 2-opt solver to the initial Nearest Neighbour solution. Note that the changes are all made in the area around Europe. On the small-scale problems examined so far, the Elves conclude that the Hybrid solver would be best suited to solve Santa's problem.

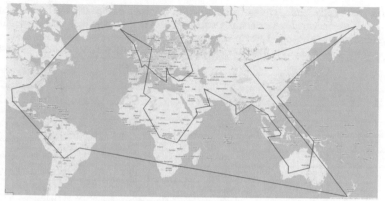

(a) A sample solution produced using the Nearest Neighbour heuristic.

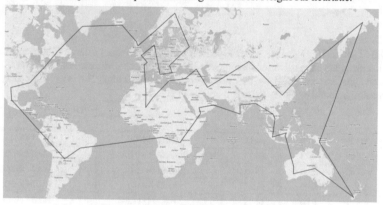

(b) A sample solution produced using the 2-Opt heuristic.

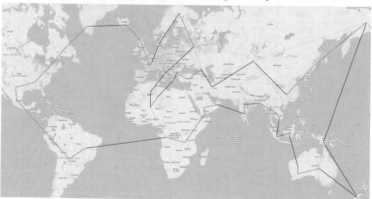

(c) A sample solution produced using the hybrid solver, overlaid onto the initial two-opt solution.

Fig. 1.6 Results produced using NN, 2-opt and the Hybrid heuristic. The route sections in brown are common to both solutions, those in gold are only in the Nearest Neighbour solution and those in blue are changes made by 2-opt as part of the Hybrid solver. *Base map and data from OpenStreetMap and OpenStreetMap Foundation* (https://www.openstreetmap.org/copyright)

Table 1.2 The results obtained for test instances with visits in the range of 14–50. The Exhaustive search is no longer included as the run times were excessive for instances with >13 visits. Note that times (milliseconds) are averaged over 10 runs and distances are presented as <min(average)> over 10 runs. As we are starting from a fixed point, NN is not stochastic and so no average is presented

Visits	NearestN		TwoOpt	
	Time	Dist	Time	Dist
14	0.00	665.11	0.3	530.78 (571.55)
15	0.00	677.64	0.3	543.2 (566.39)
20	0.00	815.5	0.8	676.26 (703.19)
25	0.00	887.28	1.8	725.15 (764.83)
30	0.00	946.7	4.9	815.52 (868.17)
35	0.00	883.8	9.7	834.98 (889.38)
40	0.00	1032.79	18.3	930.99 (963.47)
45	0.00	1064.67	24.7	934.25 (993.1)
50	0.00	1236.75	46.1	976.81 (1046.69)

Table 1.3 The results obtained using the Hybrid algorithm. The values in he Optimum column were calculated using the exhaustive search heuristic. Note that times (milliseconds) are averaged over 10 runs and distances are presented as <min(average)> over 10 runs. The items in **bold** highlight the best solution found using the three solvers under comparison

Visits	Opt.	NearestN		TwoOpt		Hybrid	
		Time	Dist	Time	Dist	Time	Dist
1	64.11	0.00	**64.11**	0	**64.11 (64.11)**	0	**64.11 (64.11)**
2	68.93	0.00	**68.93**	0	**68.93 (68.93)**	0	**68.93 (68.93)**
3	69.02	0.00	**69.02**	0	**69.02 (69.14)**	0.1	**69.02 (69.02)**
4	93.23	0.00	**93.23**	0	**93.23 (93.87)**	0.1	**93.23 (93.23)**
5	108.77	0.00	112.35	0.1	**108.77 (111.04)**	0.1	**108.77 (108.77)**
6	125.68	0.00	151.8	0.1	**125.68 (148.53)**	0	**125.68 (125.68)**
7	169.27	0.00	169.27	0	**169.27 (193.92)**	0	**169.27 (169.27)**
8	195.47	0.00	**195.47**	0	**195.47 (217.22)**	0	**195.47 (195.47)**
9	249.85	0.00	254.66	0	269.58 (288.79)	0	**249.85 (249.85)**
10	328.25	0.00	**328.77**	0	**328.77 (356.88)**	0.1	**328.77 (328.77)**
11	389.17	0.00	**389.17**	0.2	**389.17 (418.87)**	0	**389.17 (389.17)**
12	389.23	0.00	484.15	0.1	405.76 (438.6)	0.1	**401.55 (401.55)**
13	514.51	0.00	661.8	0.1	**514.51 (556.84)**	0.1	**514.51 (514.51)**
14		0.00	665.11	0.3	530.78 (571.55)	0.1	**521.86 (521.86)**
15		0.00	677.64	0.3	543.2 (566.39)	0.1	**534.4 (534.4)**
20		0.00	815.5	0.8	676.26 (703.19)	0.2	**664.34 (664.34)**
25		0.00	887.28	1.8	**725.15 (764.83)**	0.6	**725.15 (725.15)**
30		0.00	946.7	4.9	815.52 (868.17)	1	**833.44 (833.44)**
35		0.00	883.8	9.7	**834.98 (889.38)**	0.8	867.89 (867.89)
40		0.00	1032.79	18.3	930.99 (963.47)	1.5	**919.61 (919.61)**
45		0.00	1064.67	24.7	934.25 (993.1)	2.7	**925.88 (925.88)**
50		0.00	1236.75	46.1	976.81 (1046.69)	8.1	**950.49 (950.49)**

```
1    public void solve() {
2        //Create a nearest neighbour solution
3        TSPSolver nn = new NearestNTSPSolver();
4        nn.setProblem(this.theProblem);
5        nn.solve();
6
7        //Improve the NN solution using 2-opt
8        TSPSolver twoOpt = new TwoOptTSPSolver();
9        twoOpt.setProblem(this.theProblem);
10       twoOpt.solve();
```

Listing 1.4   The hybrid algorithm

Santa Claus examines the results presented and agrees with the initial conclusions that the exhaustive search is impractical for anything other than trivial problems and that the hybrid solver represents the best overall performance.

1.6   Conclusions

In this chapter, we have introduced the fundamental concepts of problem instances and problem solvers. We have examined a number of problem-solving techniques including exhaustive search, greedy heuristics and iterative improvement. If you examine recent research into the TSP (Cook 2012), then you will become aware that the examples examined in this chapter are very small indeed. The principals employed in this chapter may be applied to larger instances, but heuristics such as Lin–Kernighan (Kernighan and Lin 1973) are required in order to gain reasonable results. The reader is encouraged to download and use the Concorde (Cook et al. 2020) TSP solver which demonstrates the solving of large instances very quickly.

Within the examples covered in this book, the problems are constrained in size by the real-world considerations such as the capacity of a delivery vehicle or the length of time that a food product may be in transit.

References

Bocks, F. 1958. *An Algorithm for Solving Travelling-salesman and Related Network Optimisation Problems*. Fourteenth National Meeting: Operations Research Society of America.

Brady, R.M. 1985. Optimization Strategies Gleaned from Biological Evolution. *Nature* 317 (6040): 804–806. Bandiera_abtest: a Cg_type: Nature Research Journals Number: 6040 Primary_atype: Research Publisher: Nature Publishing Group. https://www.nature.com/articles/317804a0.

Cook, W. 2012. *In Pursuit of the Traveling Salesman: Mathematics at the Limits of Computation*. Princeton: Princeton University Press.

Cook, W., D. Applegate, R.E. Bixby, and V. Chvátal. 2020. Concorde TSP Solver. http://www.math.uwaterloo.ca/tsp/concorde.html.

Croes, G.A. 1958. A Method for Solving Traveling-Salesman Problems. *Operations Research* 6 (6): 791–812.

Cummings, Nigel. 2000. A Brief History of the Travelling Salesman Problem. *Inside OR*.

Dantzig, G.B., and J.H. Ramser. 1959. The Truck Dispatching Problem. *Management Science* 6 (1): 80–91. https://doi.org/10.1287/mnsc.6.1.80.

Dantzig, G., R. Fulkerson, and S. Johnson. 1954. Solution of a Large-Scale Traveling-Salesman Problem. *Journal of the Operations Research Society of America* 2 (4): 393–410. Publisher: INFORMS. http://www.jstor.org/stable/166695.

Flood, M.M. 1956. The Traveling-Salesman Problem. *Operations Research* 4 (1): 61–75.

Freisleben, B., and P. Merz. 1996. New Genetic Local Search Operators for the Traveling Salesman Problem. In *Parallel Problem Solving from Nature — PPSN IV*, Lecture Notes in Computer Science, ed. H.-M. Voigt, W. Ebeling, I. Rechenberg, and H.-P. Schwefel, 890–899. Heidelberg: Springer.

Johnson, D.S. 1990. Local optimization and the Traveling Salesman Problem. In *Automata, Languages and Programming*, Lecture Notes in Computer Science, ed. M.S. Paterson, 446–461. Heidelberg: Springer.

Kernighan, B.W., and S. Lin. 1973. Heuristic Solution of a Signal Design Optimization Problem. *Bell System Technical Journal* 52 (7): 1145–1159.

Knuth, D.E. 1973. *The Art of Computer Programming*. Boston: Addison-Wesley.

Menger, Karl. 1932. *Das botenproblem, Ergebnisse eines Mathematischen Kolloquiums*, vol. 2, 11–12. Leipzig: Teubner.

Mukhopadhyay, A., D. Whitley, and R. Tinós. 2021. An Efficient Implementation of Iterative Partial Transcription for the Traveling Salesman Problem. In *Proceedings of the Genetic and Evolutionary Computation Conference*, GECCO '21, Association for Computing Machinery, New York, NY, USA, 252–260. https://doi.org/10.1145/3449639.3459368

Varadarajan, S., and D. Whitley. 2021. A Parallel Ensemble Genetic Algorithm for the Traveling Salesman Problem. In *Proceedings of the Genetic and Evolutionary Computation Conference*, GECCO '21, Association for Computing Machinery, New York, NY, USA, 636–643. https://doi.org/10.1145/3449639.3459281.

Varadarajan, S., D. Whitley, and G. Ochoa. 2020. Why Many Travelling Salesman Problem Instances are Easier Than You Think. In *Proceedings of the 2020 Genetic and Evolutionary Computation Conference*, GECCO '20, Association for Computing Machinery, New York, NY, USA, 254–262. https://doi.org/10.1145/3377930.3390145.

Villanueva, J.C. 2018. How Many Atoms Are There in the Universe? *Universe Today*. https://www.universetoday.com/36302/atoms-in-the-universe/.

Watson, J., C. Ross, V. Eisele, J. Denton, J. Bins, C. Guerra, D. Whitley, and A. Howe. 1998. The Traveling Salesrep Problem, Edge Assembly Crossover, and 2-opt. In *Parallel Problem Solving from Nature - PPSN V*, Lecture Notes in Computer Science, ed. A.E. Eiben, T. Bäck, M. Schoenauer, and H.-P. Schwefel, 823–832. Berlin: Springer.

Chapter 2
Heuristics for the Vehicle Routing Problem

Abstract The Vehicle Routing Problem (VRP) represents a scenario in which multiple routes, each served by a vehicle, must be constructed in order to visit a set of customers. The VRP has two main elements to it, the allocation of visits to routes and the ordering of the visits within each route. The VRP has been investigated by many researchers and there exist many variants which may incorporate constraints such as vehicle capacities, time windows and vehicle availability. In this chapter, we discuss the solving of the CVRP using the Grand Tour and Clarke–Wright heuristics through a case study based. Our case study (*ACE Doughnuts*) describes a simple scenario where a product has to be shipped from a supplier to customers, subject to the capacity constraints of the vehicles used. In common, for the Travelling Salesperson Problem (discussed earlier in Chap. 1), the results indicate that the VRP does not scale well. We introduce the *Grand Tour* and *Clarke–Wright* heuristics and discuss the results found. It may be argued that the VRP underpins many real-world logistics problems and that the ability to understand and solve them is crucial to being able to solve more complex real-world problems.

2.1 A Description of the Vehicle Routing Problem

The VRP was first discussed by Dantzig and Fulkerson (1954) and Dantzig and Ramser (1959) where it was referred to as the "Truck Dispatching Problem". Dantzig and Ramser's problem concerned the dispatching of trucks carrying gasoline from a central terminal to service stations and was solved by using Integer Linear Programming (ILP). For an overview of the history of VRP and its many variants, the reader is directed to Gendreau et al. (2008) and Laporte et al. (2013). There are a great many variants of the VRP, including the Vehicle Routing Problem with Time Windows (VRPTW) (Solomon 1987), those with heterogeneous vehicle fleets (Jiang et al. 2014) as well as those focused upon the delivery of services (people) (Castillo-Salazar et al. 2012; Hiermann et al. 2013) rather than goods.

We will initially consider the *Capacitated Vehicle Routing Problem* (CVRP) (Ralphs et al. 2003). In this problem, we have a set of customers each of which requires a visit from a vehicle based at a central depot. Unlike TSP which requires

© Springer Nature Switzerland AG 2022

N. Urquhart, *Nature Inspired Optimisation for Delivery Problems*,

Natural Computing Series, https://doi.org/10.1007/978-3-030-98108-2_2

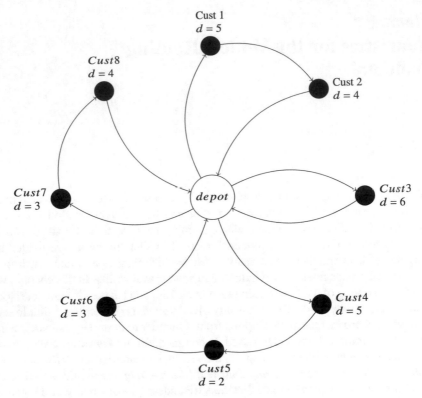

Fig. 2.1 An example of a capacitated vehicle routing problem solution. In this case, there are 8 customers with a total demand (d) of 31. The vehicles used have a capacity (cap) of 10. A solution based on 4 vehicles is shown. The minimum number of vehicles needed is $\frac{d}{cap} = 3.1$ which is rounded up to 4 vehicles

the construction of a single route, VRPs require a set of routes radiating from the depot out to customers. In the CVRP, each customer has a set *demand*, and the sum of customer demands on a route may not exceed the *capacity* of the vehicle.

Our initial examination of CVRP uses a model that has two solution characteristics; the number of vehicles and the distance travelled. Other formulations of the VRP may use characteristics such as financial costs, environmental impact or delivery times.

Figure 2.1 shows a valid solution to a small CVRP; in this case, 8 customers are served by 4 routes, each starting and ending its route at the central depot.

Two types of decisions are required when solving a VRP; firstly, the decision to allocate a particular customer to a van and secondly, the order in which a van should visit the customers.

2.2 Solving the CVRP

Let us consider the following scenario which represents a typical real-world problem:

ACE Doughnuts bake doughnuts at their factory. Customers order doughnuts, placing their order the previous day. The company has a fleet of vans, each of which can carry up to *capacity* doughnuts. Each day the company has to work out a plan that allocates deliveries to vans. The plan must ensure that:

- No van is carrying more than *capacity* doughnuts
- Each customer receives the number of doughnuts that they ordered
- The number of vans used is minimised
- The total distance travelled collectively by the vans is minimised.

Problem Instances

In order to understand the CVRP, we need a pool of known problem instances to solve; in this case, we will use the instances proposed by Augerat (1995, 2014a, b, c). The 60 CVRP instances provided by Augerat are presented with solutions arrived at using a branch and cut algorithm. Of the instances, some are presented with an optimal solution and some with a best-known solution. Tables 2.1, 2.2 and 2.3 show the problem details, n being the number of visits in each problem. Usefully, the Augerat problems are supplied with optimum (or best-known) solution values. The number of visits varies from 15 to 100; with an average of 51 visits per problem instance, the instances are divided into three sets:

- Set A—random instances.
- Set B—customers located clustered.
- Set P—instances based on previous literature.

Examples of problems from each of the sets may be seen in Fig. 2.2. For the purposes of this chapter, we will assume that the Augerat problem instances represent 60 daily sets of customer orders for the ACE Doughnut factory.

The Grand Tour

Our first attempt to solve the CVRP builds upon our attempts to solve the TSP in the last chapter. We can consider the CVRP to be a TSP, that is broken down into separate routes. Consider the example shown in Fig. 2.1. If we treat the problem as a TSP, with the depot s as the start/finish point and the customers as cities, then we might produce the following TSP solution:

$$d \rightarrow c1 \rightarrow c2 \rightarrow c3 \rightarrow c4 \rightarrow c5 \rightarrow c6 \rightarrow c7 \rightarrow c8 \rightarrow d$$

Table 2.1 The problem instances (group A) first published by Augerat (1995), with the published results shown alongside those obtained with the Grand Tour and Clarke–Wright algorithms. The number of visits within each problem is n

Problem	n	Branch and cut result			Grand Tour result		Clarke and Wright	
		Best or Optimal	Distance	Vehicles	Distance	Vehicles	Distance	Vehicles
A-n32-k5	31	Optimal	784	5	1023.89926	5	810.8470159	5
A-n33-k5	32	Optimal	661	5	777.8478133	5	712.047821	5
A-n33-k6	32	Optimal	742	6	880.5289155	6	776.2559893	7
A-n34-k5	33	Optimal	778	5	971.662654	6	810.4113356	6
A-n36-k5	35	Optimal	799	5	975.5365886	5	834.3556307	5
A-n37-k5	36	Optimal	669	5	968.2097553	5	707.2997834	5
A-n37-k6	36	Optimal	949	6	1227.87061	7	976.606366	6
A-n38-k5	37	Optimal	730	5	966.0795296	6	772.9437332	6
A-n39-k5	38	Optimal	822	5	1061.980508	5	894.7454738	5
A-n39-k6	38	Optimal	831	6	1055.150119	6	880.295466	6
A-n44-k7	43	Optimal	937	6	1085.704298	6	976.0377358	7
A-n45-k6	44	Optimal	944	6	1160.687832	7	1013.282571	7
A-n45-k7	44	Best	1146	7	1432.057072	7	1199.975963	7
A-n46-k7	45	Optimal	914	7	1159.091159	7	939.7435419	7
A-n48-k7	48	Best	1073	7	1293.15	7	1112.82	7
A-n53-k7	53	Optimal	1010	7	1208.42	8	1097.58	8
A-n54-k7	54	Best	1167	7	1694.13	8	1204.25	7
A-n55-k9	55	Optimal	1073	9	1359.22	10	1110.57	9
A-n60-k9	60	Best	1408	9	1612.34	10	1399.89	9
A-n61-k9	61	Best	1035	9	1250.01	10	1108.28	10
A-n62-k8	62	Best	1290	8	1597.82	8	1352.81	8
A-n63-k9	63	Best	1634	9	2017.38	10	1687.96	10
A-n63-k10	63	Best	1315	10	1524.77	10	1377.87	10
A-n64-k9	64	Best	1402	9	1685.17	10	1468.90	10
A-n65-k9	65	Best	1177	9	1428.03	10	1245.35	10
A-n69-k9	69	Best	1168	9	1394.78	9	1222.32	9
A-n80-k10	80	Best	1764	10	2055.26	10	1875.03	10

Table 2.2 The problem instances (Group B) first published by Augerat (1995), with the published results shown alongside those obtained with the Grand Tour and Clarke–Wright algorithms. The number of visits within each problem is n

Problem	n	Branch and cut result			Grand Tour result		Clarke and Wright	
		Best or Optimal	Distance	Vehicles	Distance	Vehicles	Distance	Vehicles
B-n31-k5	30	Optimal	672	5	722.26	5	681.16	5
B-n34-k5	33	Optimal	788	5	847.43	5	798.71	5
B-n35-k5	34	Optimal	955	5	1033.82	5	981.47	5
B-n38-k6	37	Optimal	805	6	908.47	6	831.87	6
B-n39-k5	38	Optimal	549	5	655.31	5	566.71	5
B-n41-k6	40	Optimal	829	6	1168.34	7	899.90	7
B-n43-k6	42	Optimal	742	6	809.53	6	780.80	6
B-n44-k7	43	Optimal	909	7	1189.29	8	937.74	7
B-n45-k5	44	Optimal	751	5	842.08	6	767.31	5
B-n45-k6	44	Optimal	678	6	788.58	7	734.30	7
B-n50-k7	49	Optimal	741	7	1011.12	8	749.56	7
B-n50-k8	49	Best	1313	8	1668.70	9	1354.03	8
B-n51-k7	50	Optimal	1032	7	1206.06	8	1123.90	8
B-n52-k7	51	Optimal	747	7	924.89	7	768.11	7
B-n56-k7	55	Optimal	707	7	808.78	7	739.00	7
B-n57-k7	56	Optimal	1153	7	1542.21	8	1239.78	8
B-n57-k9	56	Best	1598	9	1763.16	9	1655.75	9
B-n63-k10	62	Best	1537	10	1838.68	11	1601.31	10
B-n64-k9	63	Optimal	861	9	1126.17	10	922.19	10
B-n66-k9	65	Best	1374	9	1568.02	10	1418.13	10
B-n67-k10	66	Best	1033	10	1305.43	11	1101.71	11
B-n68-k9	67	Best	1304	9	1488.36	10	1318.41	9
B-n78-k10	77	Best	1266	10	1531.35	11	1271.90	10

This tour that takes in all of the customers is sometimes known as the *Grand Tour*. We can divide the Grand Tour into feasible sub-tours (with respect to the vehicle capacity constraint) by commencing with an empty tour:

$$d \to d$$

and adding customers until the combined demand of the customers exceeds the capacity of the vehicle. In this case, Customer 1 is added, then Customer 2 (giving a total demand of 9) and Customer 3 cannot be added as the combined demand of 15 exceeds the capacity (10) of the vehicle.

$$d \to c1 \to c2 \to d$$

Table 2.3 The problem instances (Group P) first published by Augerat (1995), with the published results shown alongside those obtained with the Grand Tour and Clarke–Wright algorithms. The number of visits within each problem is n

Problem	n	Branch and cut result			Grand Tour result		Clarke and Wright	
		Best or Optimal	Distance	Vehicles	Distance	Vehicles	Distance	Vehicles
P-n16-k8	15	Best	435	8	497.27	9	478.77	9
P-n19-k2	18	Best	212	2	278.07	3	220.64	3
P-n20-k2	19	Best	220	2	286.24	3	223.56	2
P-n21-k2	20	Best	211	2	262.82	2	225.75	2
P-n22-k2	21	Best	216	2	288.59	2	233.12	2
P-n22-k8	21	Best	603	8	682.38	10	590.62	9
P-n23-k8	22	Best	554	8	671.64	11	539.48	9
P-n40-k5	39	Best	458	5	624.15	5	524.51	5
P-n45-k5	44	Best	510	5	658.11	5	536.65	5
P-n50-k7	49	Best	554	7	685.68	7	607.33	7
P-n50-k8	49	Best	649	8	745.51	9	665.59	9
P-n50-k10	49	Best	696	10	853.71	11	734.32	11
P-n51-k10	50	Best	745	10	963.99	11	774.45	11
P-n55-k7	54	Best	524	7	755.11	7	593.76	7
P-n55-k8	54	Best	576	8	728.05	8	618.16	7
P-n55-k10	54	Best	669	10	852.56	10	736.86	11
P-n55-k15	54	Best	856	15	1074.85	18	984.08	17
P-n60-k10	59	Best	706	10	949.57	11	796.32	10
P-n60-k15	59	Best	905	15	1153.66	16	1016.96	16
P-n65-k10	64	Best	792	10	1030.82	11	851.67	10
P-n70-k10	69	Best	834	10	1013.20	11	893.81	11
P-n76-k4	75	Best	589	4	755.67	4	668.15	5
P-n76-k5	75	Best	631	5	812.75	6	706.29	6
P-n101-k4	100	Optimal	681	4	945.33	4	767.83	5

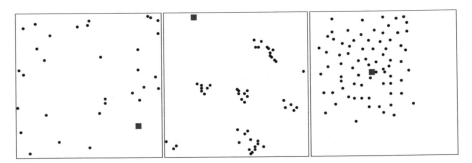

Fig. 2.2 Examples of problem instances from each of the Augerat sets (Left to right: A-n32-K5, B-n57-k9 and P-n76-k4)

A second tour commencing with Customer 3 is created, as the demand of Customers 3 and 4 would exceed the capacity of the vehicle and so tour 2 only visits one customer.

$$d \to c1 \to c2 \to d$$

$$d \to c3 \to d$$

This process continues until all of the customers have been added to a tour as follows (see Fig. 2.1).

$$d \to c1 \to c2 \to d$$

$$d \to c3 \to d$$

$$d \to c4 \to c5 \to c6 \to c7 \to d$$

$$d \to c8 \to d$$

Splitting up a Grand Tour in this manner has the advantage of producing a very quick solution, but the solution quality is very much dependent on having first solved a TSP in order to create the Grand Tour.

Algorithm 6 Creating a CVRP solution from a Grand Tour

1: **Procedure** $cvrpGrandTour()$
2: $currentTour = []$
3: $currentDemand = 0$
4: **for** $visiting rand Tour$ **do**
5:    **if** $(currentDemand + visit.demand) > CAPACITY$ **then**
6:       $result.add(currentTour)$
7:       $currentTour = []$
8:       $currentDemand = 0$
9:    $currentTour.append(visit)$
10:    $currentDemand = +visit.demand$
11: **Return** $result$
12: **EndProcedure**

The Clarke–Wright Savings Algorithm

Clarke and Wright (1964) proposed an algorithm for solving the CVRP based on the savings made by combining tours. The algorithm is an iterative algorithm that commences from a starting point based on each customer having their own tour (see Fig. 2.3). This starting point represents a poor solution to the problem. This solution

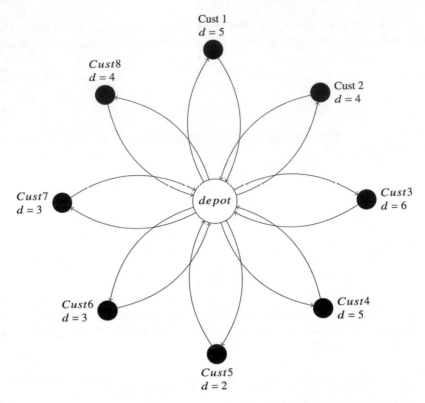

Fig. 2.3 The starting point (1 vehicle per customer) of the Clarke–Wright algorithm

can be improved by finding pairs of customers that can be placed next to each other within a tour, resulting in fewer tours and less overall distance travelled.

Each possible pair of customers is considered and the potential saving of having them in the same tour is calculated. Firstly, the distance d between them via the depot is calculated:

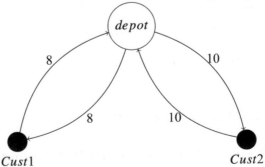

$$d = distance(c_1, d) + distance(d, c_2)$$

$$d = 8 + 10$$

$$d = 18$$

Then the direct distance $'d$ between them is calculated:

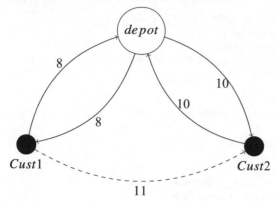

$$d' = distance(c_1, c_2)$$

$$d' = 11$$

Finally, the saving s is calculated as follows:

$$s = d - d'$$

$$s = 18 - 11$$

$$s = 7$$

As long as s is greater than 0, then there is a saving to be made (we will check the vehicle capacity constraint at a later stage). Once each pair of customers has been assessed, the savings are ordered such that the pairing that gives the biggest saving is at the head of the list.

The algorithm now considers each saving in the order that it is stored within the list, and this ensures that the savings that have the biggest benefits are considered first which maximises their chances of being successfully applied to the solution.

If we consider a saving that places c_x next to c_y, we need to establish that the following conditions are met before the saving may be applied:

- c_x is at the end of a route.
- c_y is at the start of a route.

- The combined demand of the routes that incorporate c_x and c_y is less than $vcap$.

The complete Clarke and Wright algorithm is shown in listing Algorithm 7, lines 2–10 calculate the savings and lines 11–18 sort and apply them to the solution.

Algorithm 7 The Clarke–Wright Algorithm

1: **Procedure** $ClarkeWright()$
2: $savings = []$
3: $solution = buildOnePerCust()$
4: **for** $visit v_x in problem$ **do**
5:     **for** $visit v_y in problem$ **do**
6:         $d = distance(v_x, depot) + distance(depot, v_y)$
7:         $d' = distance(v_x, v_y)$
8:         $s = d - d'$
9:         **if** $s > 0$ **then**
10:            $savings.add(new(saving(s, v_x, v_y))$
11: $sort(savings)$
12: **for** $savings in savings$ **do**
13:     $r_x = route(s.x)$
14:     $r_y = route(s.y)$
15:     **if** $r_x.last == s.x \&\& r_y.first == s.y$ **then**
16:         **if** $d(r_x) + d(r_y) <= cveh$ **then**
17:             $merge(r_x, r_y)$
18: **Return** $solution$
19: **EndProcedure**

2.3   Implementation

Our Java implementation of the CVRP solvers is based upon our earlier attempts to solve the TSP (see Fig. 1.4), and the updated class diagram is shown in Fig. 2.4. We make use of the same patterns used when solving the TSP, remembering to separate the problem from the solver. All of the code used can be found in the repository associated with this book.

Within the *CVRPProblem* class, we store the list of visits as an *ArrayList* of *VRPVisit* objects, and this represents the unsolved problem. The solution (as created by the solver) comprises an *ArrayList* object which contains the set of tours. Each tour is represented by an *ArrayList* of *VRPVisit* objects.

Our *GrandTour* implementation is shown in Listing 2.1, and it assumes that the list of visits has already been ordered using a TSP operator (see Listing 2.2 which makes use of the *NearestNeighbour* TSP solver as discussed in Chap. 1). The *GrandTour* solver follows the pseudocode closely; having only one loop and an if statement, it is of relatively low complexity and therefore produces a solution quickly, although as we shall see in the next section the quality of the solution produced is poor compared to other methods.

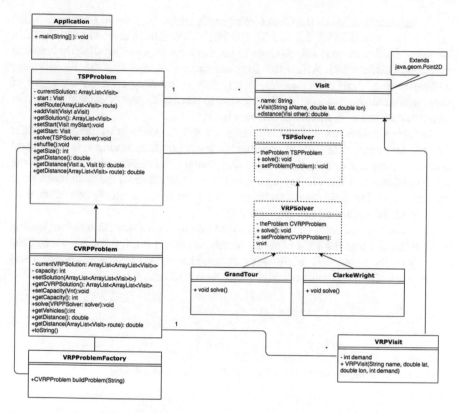

Fig. 2.4 The VRP solver architecture

```
1    public void solve() {
2        ArrayList<ArrayList<VRPVisit>> solution = new ArrayList<
         ArrayList<VRPVisit>> ();
3        ArrayList<VRPVisit> currentRoute = new ArrayList<
         VRPVisit>();
4
5        for (Visit v : super.theProblem.getSolution()){
6            VRPVisit currentVisit = (VRPVisit)v;
7            //Can current visit be appended to the current route
         ?
8            if (currentVisit.getDemand() + routeDemand(
         currentRoute) >  this.theProblem.getCapacity()){
9                //Add current route to solution and begin a new
         route
10               solution.add(currentRoute);
11               currentRoute = new ArrayList<VRPVisit>();
12           }
13           //Add visit to the vurrent route
14           currentRoute.add(currentVisit);
15       }
16       solution.add(currentRoute);
17       super.theProblem.setSolution(solution);
18   }
```

Listing 2.1 The Grand Tour algorithm

The implementation of the Clarke–Wright algorithm is given in Listings 2.3, 2.4 and 2.5. Initially (Listing 2.3:line 3), the initial solution is created (see Fig. 2.3). We then identify the possible savings to be made by proceeding directly between customers (Listing 2.4). Within this formulation, we assume that the distances are symmetric (e.g. $dist(x, y) == dist(y, x)$); hence, we only consider each pair of customers once. We use an inner class *Saving* to store each pair of customers and the saving associated with travelling between them.

Having calculated each possible saving, we sort the savings so that we consider the greatest saving first (Listing 2.3 line 7). Each potential saving s is considered in turn (Listing 2.5 line 3). The routes that contain the customers $s.x$ and $s.y$ are identified (lines 8–130). The saving can only be implemented if the customers are at the start and end of their routes, and if either of them is in the middle of a route, then they cannot be connected directly (line 14).

The method *joinRoutes* (line 15) is used to join the two routes, and before joining them, it checks that the total capacity of the combined route does not exceed the capacity of the vehicle. If *joinRoutes* can combine the two routes, it does so and returns *true*, and if it cannot combine them, then it returns *false*.

```
1   NearestNTSPSolver nn = new NearestNTSPSolver();
2   ((TSPProblem)myVRP).solve(nn);
3   GrandTour gt = new GrandTour();
4   myVRP.solve(gt);
5   System.out.print("GTDist," + myVRP.getDistance());
6   System.out.print(",GTVehicles," + myVRP.getVehicles());
```

Listing 2.2   Using the GrandTour solver initialised with a TSP solution

```
1    public void solve() {
2        //1.Create the initial solution (1 vehicle per customer)
3        ArrayList<ArrayList<VRPVisit>> solution =
     createDefaultSolution();
4        //2.Find Savings
5        ArrayList<Saving> savings = findSavings();
6        //3.Sort the savings into the order of largest saving
     first
7        Collections.sort(savings);
8        applySavinngs(solution, savings);
9        //3.Check solution
10       checkSolution(solution);
11       //4.Adopt solution
12       super.theProblem.setSolution(solution);
13   }
```

Listing 2.3   The Clarke and Wright algorithm: the main outline

```
1    private ArrayList<Saving> findSavings() {
2        ArrayList<Saving> savings = new ArrayList<Saving>();//
     Store savings in this list
3        //Check the savings to be made for each pair of
     customers
4        for (int countX =0; countX < super.theProblem.getSize()
     ; countX ++){
5            for (int countY = countX +1; countY < super.
     theProblem.getSize(); countY++){
6                VRPVisit x = (VRPVisit)super.theProblem.
     getVisits().get(countX);
7                VRPVisit y = (VRPVisit)super.theProblem.
     getVisits().get(countY);
8                double cost = super.theProblem.getDistance(x,
     super.theProblem.getStart()) +super.theProblem.getDistance(
     y, super.theProblem.getStart());
9                double direct = super.theProblem.getDistance(x,y
     );
10               double saving = cost - direct;
11               savings.add(new Saving(x,y,saving));
12           }
13       }
14       return savings;
15   }
```

Listing 2.4 The Clarke and Wright algorithm: finding the savings

2.4 Results

ACE Doughnuts solve their collection of 60 problems using the Grand Tour and Clarke–Wright solvers, and the results can be seen in Tables 2.1, 2.2 and 2.3. To place these results into context, we also show the results obtained by Augerat using a branch and cut algorithm (Augerat 1995).

The results obtained by our two solvers are disappointing; the distances are significantly greater than those obtained by Augerat using branch and cut. For the A, B and P sets, the Grand Tour algorithm is on average 25, 20 and 27% longer and frequently has more routes in the solution. The Clarke–Wright algorithm performs better (the solutions are on average 5, 4 and 8% longer than those presented by Augerat), but it still fails to find the optimum, or even the best, results noted by Augerat. If we select a problem instance at random (P-n20-k2, Fig. 2.5a), we can see it solved as a TSP using the NN heuristic in Fig. 2.5b. If we apply the Grand Tour algorithm, that results in the solution shown in Fig. 2.5c. Note that the CVRP solution retains many of the imperfections from the TSP solution such as the loop around the visits in the lower left corner. Note also that the GT solution also comprises 3 routes, when the best solution found by Augerat (1995) and the Clarke–Wright solution (see Table 2.3) both of which only comprise 2 routes.

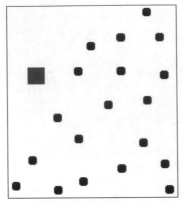

(a) The P-n20-K2 problem instance.

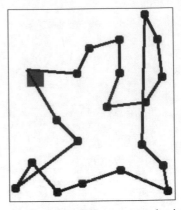

(b) The initial TSP tour created using Nearest Neighbour.

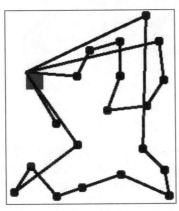

(c) The tour shown in 2.5b split into CVRP tours using the Grand Tour algorithm.

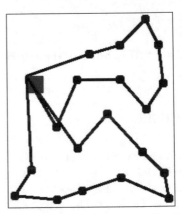

(d) The same problem solved using the Clarke-Wright algorithm.

Fig. 2.5 The P-n20-K2 problem instance, solved by creating a Grand Tour using Nearest Neighbour and then splitting into sub-tours and solved using the Clarke-Wright algorithm

2.5   Conclusions

What can we learn from this? Firstly, although the Grand Tour algorithm is simple and relatively quick, it does not produce a particularly good result. The Grand Tour takes a TSP solution as its input, and that solution has been optimised only on distance. When the Grand Tour is split into individual routes, customers can only be adjacent to those that they were adjacent to in the TSP solution or adjacent to a depot. In order to take into the more complex CVRP problem, it may be necessary to group customers together who would not be adjacent in a TSP solution.

There is nothing in the results obtained using the Grand Tour or Clarke–Wright algorithms that would want to make ACE Doughnuts adopt either of them as a replacement for branch and cut. The relevance of these results is that they both represent methods of finding *feasible* solutions to the problem in a very short space of time. Grand Tour and Clarke–Wright have an advantage over branch and cut in that they are easy to implement and understand compared to the complex linear programming techniques used in Branch and Cut, this being a major practical advantage for developers. In order to achieve results comparable with Branch and Cut, we require a more powerful technique, but one which is easily implemented.

```
1   private void applySavinngs(ArrayList<ArrayList<VRPVisit>>
        solution, ArrayList<Saving> savings) {
2       //Search through the savings and see which can be
    implemented
3       for (Saving s : savings){
4           ArrayList<VRPVisit> routeX = null, routeY =null;
5           boolean found = false;
6           //Find the routes that contain the customers
7           //at the start/end
8           for (ArrayList<VRPVisit> route : solution){
9               if (end(route)==s.x)
10                  routeX = route;
11              if (start(route)== s.y)
12                  routeY = route;
13          }
14          if ((routeX != null)&&(routeY != null)){
15              found = joinRoutes(solution, routeX, routeY);
16          }
17          if (!found){//Check for the reverse (y to x)
18              for (ArrayList<VRPVisit> route : solution){
19                  if (end(route)==s.y)
20                      routeX = route;
21                  if (start(route)== s.x)
22                      routeY = route;
23              }
24              if ((routeX != null)&&(routeY != null)){
25                  joinRoutes(solution, routeX, routeY);
26              }
27          }
28      }
```

Listing 2.5 The Clarke and Wright algorithm: applying the savings

References

Augerat, P. 1995. Approche polyèdrale du problème de tournées de véhicules. (Polyhedral approach of the vehicle routing problem). PhD Thesis, Grenoble Institute of Technology, France. https://tel.archives-ouvertes.fr/tel-00005026.

Augerat, P. 2014a. VRP-REP: Augerat 1995 Set A. http://www.vrp-rep.org/datasets/item/2014-0000.html.

Augerat, P. 2014b. VRP-REP: Augerat 1995 Set B. http://www.vrp-rep.org/datasets/item/2014-0001.html.

Augerat, P. 2014c. VRP-REP: Augerat 1995 Set P. http://www.vrp-rep.org/datasets/item/2014-0009.html.

Castillo-Salazar, J.A., D. Landa-Silva, and R. Qu. 2012. A Survey on Workforce Scheduling and Routing Problems. In *Proceedings of the 9th International Conference on the Practice and Theory of Automated Timetabling*, 283–302.

Clarke, G., and J.W. Wright. 1964. Scheduling of Vehicles from a Central Depot to a Number of Delivery Points. *Operations Research* 12 (4): 568–581. https://ideas.repec.org/a/inm/oropre/v12y1964i4p568-581.html.

Dantzig, G.B., and D.R. Fulkerson. 1954. Minimizing the Number of Tankers to Meet a Fixed Schedule. *Naval Research Logistics Quarterly* 1 (3): 217–222. https://onlinelibrary.wiley.com/doi/pdf/10.1002/nav.3800010309. https://onlinelibrary.wiley.com/doi/abs/10.1002/nav.3800010309.

Dantzig, G.B., and J.H. Ramser. 1959. The Truck Dispatching Problem. *Management Science* 6 (1): 80–91. https://doi.org/10.1287/mnsc.6.1.80.

Gendreau, M., J.-Y. Potvin, O. Bräumlaysy, G. Hasle, and A. Løkketangen. 2008. Metaheuristics for the Vehicle Routing Problem and Its Extensions: A Categorized Bibliography. In *The Vehicle Routing Problem: Latest Advances and New Challenges*, ed. B. Golden, S. Raghavan, and E. Wasil. Operations Research/Computer Science Interfaces. Springer US, Boston, MA, 143–169. https://doi.org/10.1007/978-0-387-77778-8-7.

Gilbert Laporte, P.T. 2013. Vehicle Routing: Historical Perspective and Recent Contributions. *EURO Journal on Transportation and Logistics* 2(1–2).

Hiermann, G., M. Prandtstetter, A. Rendl, J. Puchinger, and G. Raidl. 2013. Metaheuristics for solving a multimodal home-healthcare scheduling problem. *Central European Journal of Operations Research* 23 (1).

Jiang, J., K.M. Ng, K.L. Poh, and K.M. Teo. 2014. Vehicle Routing Problem with a Heterogeneous Fleet and Time Windows. *Expert Systems with Applications* 41 (8): 3748–3760. https://www.sciencedirect.com/science/article/pii/S0957417413009494.

Ralphs, T., L. Kopman, W. Pulleyblank, and L. Trotter. 2003. On the Capacitated Vehicle Routing Problem. *Mathematical Programming* 94 (2): 343–359. https://doi.org/10.1007/s10107-002-0323-0.

Solomon, M.M. 1987. Algorithms for the Vehicle Routing and Scheduling Problems with Time Window Constraints. *Operations Research*, Vol. 35, NO. 2. (MAR.- APR., 1987) 35: 254–265. https://doi.org/10.1007/s10107-002-0323-0.

Chapter 3
Applying Evolution to Vehicle Routing Problems

Abstract Previously in Chaps. 1 and 2, the author discussed the Travelling Salesperson Problem (TSP) and the Vehicle Routing Problem (VRP) (Dantzig and Fulkerson 1954; Dantzig et al. 1954; Dantzig and Ramser 1959). We noted that both of these problems underpinned many real-world delivery problems. Having examined a number of heuristics that can produce solutions to the VRP, we noted that the results obtained were disappointing when compared to benchmark results (Augerat 2014a, b, c). In this chapter, we introduce the *Evolutionary Algorithm* (EA) and demonstrate the manner in which the EA may be used to find solutions to the VRP. The EA approach has the advantage of allowing for considerable flexibility to incorporate features such as multiple constraints and user choice, as will be explored in later chapters.

3.1 Evolutionary Computation

Evolutionary Computation (EC) (Holland 1975) describes a range of metaheuristics that may be used for problem-solving, which are inspired by Darwinian Evolution (Darwin 1859). These approaches are based around the concept of *evolving* a solution to a problem by applying random changes to a *population* of solutions over time. New solutions are created within the population by applying random changes, where these changes result in an improvement that there is an increased possibility of that solution remaining in the population replacing a solution of poorer quality. In this chapter, we examine the application of *Evolutionary Algorithms* (EAs) to the CVRP.

It is important to remember that EC is *inspired* by evolution, and it is not, in most cases, an accurate biological model of evolution. For example, the EAs described in this book use a fixed population size, whereas in nature the population size of a species will ebb and flow under the influence of the environment. Some of the terminologies used within EAs (such as *genes* and *chromosomes*) have been "borrowed" from biology but their use in context of EAs does not always match the strict biological definitions of such terms. Computing can be used to model biological evolution in order to allow experimentation to take place *in silico*, but this is generally a separate field to the use of EAs for optimisation as discussed here.

© Springer Nature Switzerland AG 2022
N. Urquhart, *Nature Inspired Optimisation for Delivery Problems*,
Natural Computing Series, https://doi.org/10.1007/978-3-030-98108-2_3

As well as optimisation, Evolutionary techniques have been used across a range of applications from the evolution of artworks to the evolution of software. The reader should remember that EAs are just one algorithmic technique within the field of Natural Computation. Other techniques include neural networks (Hertz et al. 1991), artificial immune systems (Kephart 1994) and ant colony optimisation (Colorni et al. 1992) amongst many others.

3.2 Evolutionary Algorithms

The term Evolutionary Algorithm covers a wide variety of algorithms based around the evolutionary metaphor. In order to allow evolution to take place, a *fitness function* is required. The role of the fitness function is to take any possible solution (from a random solution to a carefully constructed optimal or near-optimal solution) and evaluate it. A simple fitness function might return a single value in the range from 0 to ∞, 0 representing the optimal solution and the values increasing as the solution becomes less optimal.

Let us consider a simple example; the max ones problem,, the objective is to evolve a binary string of all 1s, the fitness f being the sum of all of the 1s in the string.

For example, 11011101 has a fitness of 6, and if the string is modified to 11011100, it has a fitness of 5. Over time, the evolutionary process will promote strings with more 1s until finally a string (11111111) is evolved where $f = 8$. Evolutionary processes are driven by the fitness function, and changes to solutions that improve the value of f are more likely to be retained, whilst changes that worsen f are more likely to be discarded.

A set of solutions, known as the *population*, is maintained. Each member of the population is evaluated and its *fitness* calculated; thus, any two (or more) members of the population may be compared and the most fit solution is identified. The evolutionary process seeks to undertake the gradual improvement of the solutions within the population, by replacing those with the worst fitness values with modified individuals which have a better fitness. Over time, the collective fitness of the population improves until it reaches a level that represents an acceptable solution to the end user. The basic evolutionary cycle used in most EAs is given in Algorithm 8.

The population is initialised with a random set of solutions (line 2), and they will probably have poor fitness values at this stage. The random population does, however, have the property of being *diverse*, and members of the population can encompass many different ways of solving the problem. Using the fitness function, each member of the population may be evaluated, allowing those individuals which represent higher quality solutions to be identified, along with those that represent poorer solutions to be identified.

The evolutionary cycle now commences (lines 3–7), the first stage being the selection of *parent* solutions. The parent solutions are selected at random, but with a bias towards the selection of individuals with a better fitness. The higher the quality

---

Algorithm 8 The Evolutionary Cycle

1: **Procedure** Evolve()
2: Initialise population
3: **while** !finished **do**
4:     select parents
5:     create children
6:     evaluate children
7:     replace children into population
8: **Return** best in population
9: **EndProcedure**

---

of the solution represented by an individual, the greater the chance of being selected to reproduce. By this means, individuals which represent high-quality solutions have their traits replicated in new individuals, thus increasing the propensity of that trait within the population resulting in the increase in the overall fitness of the population. It is worth noting that the EAs in this book (and most other EAs used for optimisation applications) make use of a *steady state population* which has no concept of ageing and death. Even the most fit plant or animal will one day die, although it may produce many offsprings that continue its genetic material. In the EAs described here is no concept of ageing that results in an individual becoming less likely to reproduce as they get older and eventually dying. Individuals may be replaced within the population, but the likelihood of replacement is due to fitness rather than age.

Having selected parents, new *child* solutions are produced using a combination of operators:

- **Crossover**: (Sometimes known as recombination) Crossover takes two (or more) parents and creates a child based on features from both parents.
- **Cloning** : It creates a child based on copying a single parent.
- **Mutation**: It takes a child created by recombination or cloning and introduces a small random change.

Having created children, they are copied back into the main population, replacing existing population members, with lesser fitness values. The evolutionary cycle continues until some halting criterion has been met, and it may be after a set number of cycles, by a specific time or when a fitness goal has been reached.

Each individual within the population contains a representation of a solution, sometimes referred to as a *genotype*. The genotype could be a permutation or a binary string or some other set of data. Each genotype can be transformed into a solution or *phenotype* which can then be evaluated and the fitness calculated.

The success of Evolutionary Algorithms may be partially explained by the *Building Block Hypothesis* (Goldberg 1989), based upon the concept of the discovery and promotion of useful sequences, known as the *building blocks* within the genotype. In the case of a CVRP problem, a building block might be grouping adjacent visits together or ensuring that a vehicle is used to 100% of its capacity. An individual that contains useful building blocks should have a better fitness, and if recombination creates a child that contains blocks from the parents, this may result in the child having a greater fitness than either of the parents.

For example, a CVRP that comprises 20 deliveries to customers c_0 to c_{20} holds different permutations of customers within the population. One of the solutions contains the sequence $\ldots c_{20}, c_4, c_6, c_{18} \ldots$ within the chromosome which places a set of adjacent customers into the same route. This sequence of genes improves the fitness of the individual, increasing its chances of being selected for recombination and lessening its chances of being replaced. When children are created, there is a reasonable chance that the useful sequence of genes will be copied to the child, which will in turn have an improved fitness due to this feature. In this way, a useful building block of genes can be reproduced throughout the population. This discovery and promotion of building blocks go some way to explain the success of evolutionary algorithms when applied to combinatorial optimisation problems.

3.3 Applying Evolution to the CVRP

Let us revisit the scenario that we described in Sect. 2.2, ACE Doughnuts now decided to apply an Evolutionary Algorithm to their delivery problem. The algorithm will take a CVRP problem instance and evolve a solution that has the lowest possible travel distance. Within each iteration of the evolutionary cycle, one child will be created, evaluated and potentially replaced into the main population. The fitness of an individual will be the total travel distance of the solution that it represents. The lower the fitness, the better the quality of the individual.

The elements of the algorithm may be summarised as follows:

- **Representation**: The phenotype encodes a delivery plan using a permutation of visits which will form a Grand Tour. Each visit can be regarded as a single *gene*.
- **Fitness function**: The fitness function splits the genotype (the single Grand Tour permutation—see Sect. 2.2) into the phenotype which comprises the set of vehicle routes. The fitness is the sum of the route lengths. The conversion of a genotype to a phenotype for evaluation is sometimes referred to as *decoding*.
- **Selection Strategy**: *Tournament selection* is to be used. Tournament selection picks k individuals randomly from the population, and the individual with the best fitness of those picked is then returned.
- **Population Strategy**: The population will remain a constant size, with each child replacing one member of the population (known as a *steady state* population).
- **Crossover operator**: The crossover operator uses a form of *single point* crossover which copies a randomly selected section of the genotype from the first parent (from the start to a randomly selected point) and then copies the remaining material from the second parent.
- **Mutation Operator**: The mutation operator randomly selects a single gene and moves to a random location within the genotype. This has the effect of moving a customer visit with the delivery plan, possibly between routes.

- **Replacement Strategy**: The replacement strategy performs a inverse tournament selection and the looser is replaced by the child, assuming the child has a better fitness.
- **Halting Criterion**: The algorithm will halt after a fixed number of solutions have been evaluated.

Each of these elements will be described in more detail in the sections that follow.

Representation and Evaluation

Within our representation, the genotype is a permutation of visits, and the phenotype is the set of routes that form the final solution. The evaluation function is responsible for mapping from the genotype to the phenotype. The contents of the genotype may be thought of as a set of instructions that are *decoded* in order to build the genotype solution.

In this case, the decoding mechanism splits the Grand Tour into sub tours using the capacity constraint in the same manner as the Grand Tour described in Sect. 2.2. Decoding mechanisms may be fairly complex and can include features to ensure that the resulting phenotype represents a valid solution. In this case, the decoder ensures that the capacity constraint is always met, creating new tours if necessary.

A representation that uses a genotype which cannot be directly evaluated, requiring to be decoded into a phenotype first, is known as an *indirect representation*. Conversely, a *Evolutionary Algorithm* requires no decoding as it represents the solution. If we wished to use a direct representation then we would have to not only order the deliveries within the genotype but also specify which tour they were to be incorporated into. This could lead to infeasible solutions, i.e. one that has too many visits allocated to a route, causing it to break the vehicle capacity constraint. When an indirect representation is used, the decoding stage can be used to ensure that constraints are met in the resulting solution. The decoding stage can (as we shall see in future examples) become quite complex as it ensures that the solution it is constructing is feasible. The reader may wish to think of the phenotype within an indirect solution as containing a set of instructions which are then carried out by the decoder in order to construct the genotype containing the solution.

Having constructed the phenotype solution, the fitness f of the individual may now be calculated. In this example, f is the sum of the distance travelled across all the sub tours. As well as a fitness value, a solution may have other *characteristics* of interest to the user. A characteristic of the solutions being created for ACE Doughnuts is the number of sub tours, which will relate to the number of actual vehicles required for the deliveries. We will make more use of solution characteristics in later examples, but suffice to say that such characteristics inform us about the nature of the solution and are of interest to the end user who ultimately has to adopt the solution.

Evolutionary Operators

The crossover operator creates a new child genotype incorporating elements of both parents. The operator used selects a random position x within the genome. All of the genetic material from before x is copied from parent 1 directly to the child. The incomplete child chromosome is then completed by considering each element in parent 2 and appending it to the child if it is not already present within the chromosome.

Assuming that we have genomes of 31 genes in length (representing 31 customer deliveries), consider the following example, taken at random from a run of the EA.

```
       0   1   2   3   4  5 6 7   8   9  10 11  12 9  13  14  15  16  17  18  19  20  21  22  23  24  25  26 27 28 29 30

P1   | 27| 2 |14 | 4 |24 |3|7|8 |17 |29 | 5 |12| 9 | 19 10 23 15 18 20 32 22 13 31 25 30 16 11 26  6 21 28

P2   |27 14 4 24 3 7 8 17 2 29 5 12 9 |10 |23 |19 |15 |13 |22 |32 |18 |20 |31 |25 |30 |16 |11 |26 | 6 |21 |28|

Child| 27  2  14  4  24  3 7 8  17 29  5  12 9  10 23 19 15 13 22 32 18 20 31 25 30 16 11 26  6 21 28
```

Genomes $P1$ and $P2$ are selected from the population using a tournament of size 2, and x is set to a value of 13 (selected at random).

In the above example, the fitness values of the parents were 876.05 and 878.03, respectively, once they had been decoded and evaluated. The resulting child had a fitness of 874.88.

The example shown was taken after the EA has been running for sometime; hence, they represent solutions that are already fairly optimised. If we examine the genotypes of both parents, we see that they already contain similar material. The resulting phenotypes may be seen in Fig. 3.1a, b and c.

The mutation operator selects an element at random from within the genotype and moves it to a randomly selected position, within the genotype. It may be argued that the crossover operator exploits material that already exists within the pool of genotypes by recombining it in different ways, whilst the mutation operator creates new material within the genome.

The procedure used for creating children may be seen in Algorithm 9. Note that there is a 50% chance that the child will be created by crossover; otherwise, it is created by copying a single parent. All children have a mutation applied to them.

3.4 Choosing Parameters

A major consideration when using an Evolutionary Algorithm is the setting of algorithm parameters such as population size, mutation rate and crossover rate. In the algorithm presented, ACE Doughnuts use the the following parameters:

Fig. 3.1 An example of a child solution (**c**) being created from two parents (**a**) and (**b**)

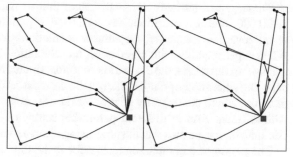

(a) The phenotype resulting from P1.

(b) The phenotype resulting from P2.

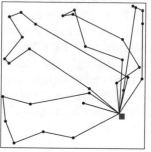

(c) The child phenotype resulting from the crossover of P1 and P2 (figures 3.1a and 3.1b).

Algorithm 9 Child Creation

1: **Procedure** $CreateChild()$
2: **if** $randBoolean() == true$ **then**
3:     $parent = tournament(2)$
4:     $child = copy(parent)$
5: **else**
6:     $parent1 = tournament(2)$
7:     $parent2 = tournament(2)$
8:     $child = crossover(parent1, parent2)$
9: $child = mutate(child)$
10: $evaluate(child)$
11: **Return** $child$
12: **EndProcedure**

- **Population Size**—1,000.
- **Evaluation Budget**—1,000,000 evaluations.
- **Crossover rate**—0.5.
- **Mutation rate**—1.
- **Selection Pressure**—tournament size of 2.
- **Replacement Pressure**—tournament size of 2.

These parameters will cause the algorithm to halt after an *evaluation budget* 1,000,000 solution evaluations has been used. If parameters such as population size or crossover rate are altered, the number of solutions evaluated remains at 1,000,000. When experimenting with algorithms, the evaluation budget is a useful variable to manage. In most EAs, the evaluation function, (including the genotype to phenotype decoding where appropriate) is likely to be the most complex part of the algorithm and therefore where the majority of processing time is spent. If we wish to experiment with different EAs or different parameter settings keeping the evaluation budget constant, then it ensures that the processing effort expended is kept roughly constant.

The crossover and mutation rates refer to the likelihood of these operators being applied to a new individual; in this case, there is a 50% probability that a child will be created as a result of crossover (the other 50% will be created by cloning), and every child will have a mutation applied to it after creation.

Setting such parameters can pose difficulties for developers such as those working for ACE Doughnuts who have limited experience with EAs. In this case, the parameters are set through trial and error. The automated selection or *tuning* of EA parameter values may be carried out by a number of methods (Eiben and Smit 2012). An in-depth discussion of parameter selection and tuning is out of the scope of this book; if the reader is interested in finding out more, they are directed to Eiben and Smit (2012).

3.5   Implementation

ACE Doughnuts implement the EA described above in Java; as with the previous Clarke–Wright implementation (see Sect. 2.3), they extend *VRPSolver* and use the existing framework.

The Evolutionary Algorithm is implemented based around a *VRPea* class that contains the evolutionary loop and an *Individual* class that contains the genotype and phenotype of one solution, along with the crossover operator, mutation operator and evaluation function. As there are many variations on the Evolutionary Algorithm, this architecture separates the evolutionary loop from the Individual solution, and this architecture allows the Individual to be re-used (Fig. 3.2).

The Evolutionary Algorithm is *stochastic* due to the random nature of operators such as selection, mutation and crossover. We make use of the *java.util.Random* class to provide random numbers during the execution of the algorithm. It is desirable to be able to repeat runs, for debugging purposes, and to allow results to be reproduced. To allow for this, a *java.util.Random* object may be initialised with a seed value which causes the same sequence of random values to be issued each time it is initialised with that seed. In our implementation of the EA, we use a class *RandomSingleton*, and this incorporates an instance of *java.util.Random* which can be seeded and is contained within an instance of the *Singleton* design pattern.

Fig. 3.2 The class *V R P ea*
extends *V R P Solver* (see
Fig. 2.4). A collection of
Individual objects are used
by *V R P ea* as the population

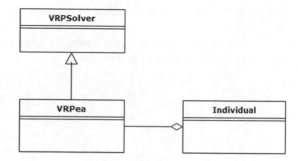

The CVRPea Class

The *CVRPea* class (Listing 3.1) contains the evolutionary loop (lines 14–42). In our
implementation, one child is created during each iteration of the loop. The population
of Individuals is contained within the ArrayList *population*, and the most fit member
of the population is kept track of using the *bestSoFar* reference. New children are
created by crossover or cloning (lines 18–25) and then mutated and evaluated. Note
that the loop is controlled by the evaluations budget, which is decremented (line 28)
each time a solution is evaluated.

The Individual Class

Our architecture encapsulates the implementation of the genotype, phenotype and
the associated operators within the *Individual* class. This allows the problem-specific
details, such as representation and operators, to be kept separate from the algorithm
located in the solver class (e.g. *VRPea*). This approach makes it easier to reuse an
algorithm for similar problems by replacing the *Individual* class or trying a different
algorithm.

Each solution within the population is an instance of the *Individual* class which
encompasses the Genotype and Phenotype data structures (Listing 3.2). The mutation
and crossover operators act on the genotype so are included within *Individual*. The
evaluate() method which constructs the phenotype by decoding the genotype is also
located within *Individual*.

The *Individual* class uses two constructors (see Listing 3.3); the first is used to
create a new Individual with a randomised genotype, and the second takes two parent
Individuals and uses the crossover operator to initialise a genotype using the scheme
outlined in Sect. 3.3. Where an Individual is created by cloning, the *copy()* method
is used to copy the genotype of the parent directly to the child.

The *Individual.evaluate()* method is defined in Listing 3.5. The *evaluate()* method
iterates through each visit in the genotype (line 10) and adds it to the current route,
until the collective demand of the route exceeds the vehicle capacity (line 11). At

that point, the current route is added to the phenotype (line 14) and a new route is created (line 15).

A full listing of each item of the Individual class can be found in the source code repository that accompanies this book.

```
1    private RandomSingleton rnd = RandomSingleton.getInstance()
     ;
2    //Note that we use the RandomSingleton object to generate
     random numbers
3
4    //EA Parameters
5    private int POP_SIZE = 500;
6    private int TOUR_SIZE = 2;
7    private double XO_RATE = 0.7;
8    private int evalsBudget = 1000000;
9
10   @Override
11   public void solve() {
12       //Reference to the best individual in the population
13       Individual bestSoFar = InitialisePopution();
14       while(evalsBudget >0) {
15           System.out.println(evalsBudget);
16           //Create child
17           Individual child = null;
18           if (rnd.getRnd().nextDouble() < XO_RATE){
19               //Create a new Individual using recombination,
     randomly selecting the parents
20               child = new Individual(super.theProblem,
     tournamentSelection(TOUR_SIZE),tournamentSelection(
     TOUR_SIZE));
21           }
22           else{
23               //Create a child by copying a single parent
24               child = tournamentSelection(TOUR_SIZE).copy();
25           }
26           child.mutate();
27           child.evaluate();
28           evalsBudget --;
29
30           //Select an Individual with a poor fitness to be
     replaced
31           Individual poor = tournamentSelectWorst(TOUR_SIZE);
32           if (poor.evaluate() > child.evaluate()){
33               //Only replace if the child is an improvement
34
35               if (child.evaluate() < bestSoFar.evaluate()){
36                   bestSoFar = child;
37               }
38               //child.check();//Check child contains a valid
     solution
39               population.remove(poor);
40               population.add(child);
41           }
42       }
43       super.theProblem.setSolution(bestSoFar.getPhenotype());
44   }
```

Listing 3.1 The evolutionary algorithm

```
1    //The genotype is a "grand tour" list of visits
2    private ArrayList<VRPVisit> genotype;
3
4    //The phenotype is a set of routes created from the
     genotype
5    private ArrayList<ArrayList<VRPVisit>> phenotype;
```

Listing 3.2 The genotype and phenotype properties within the Individual class

```
1    public Individual( CVRPProblem prob) {
2        /*
3         * Constructor to create a new random genotype
4         */
5        problem = prob;
6        genotype = new ArrayList<VRPVisit>();
7        for (Visit v : prob.getSolution()){
8            genotype.add((VRPVisit)v);
9        }
10       genotype = randomize(genotype);
11       phenotype = null;
12   }
13
14   public Individual (CVRPProblem prob, Individual parent1,
     Individual parent2){
15       /*
16        * Create a new Individual based on the recombination
     of genes from <parent1> and <parent2>
17        */
18       problem = prob;
19       genotype = new ArrayList<VRPVisit>();
20       int xPoint = rnd.getRnd().nextInt(parent1.genotype.size
     ());
21
22       //copy all of p1 to the xover point
23       for (int count =0; count < xPoint; count++ ){
24           genotype.add(parent1.genotype.get(count));
25       }
26
27       //Now add missing genes from p2
28       for (int count =0; count < parent2.genotype.size();
     count++){
29           VRPVisit v = parent2.genotype.get(count);
30           if (!genotype.contains(v)){
31               genotype.add(v);
32           }
33
34       }
35   }
```

Listing 3.3 The constructors used within the Individual class

```
1    public void mutate() {
2        //Mutate the genotype, by randomly moving a gene.
3        phenotype = null;
4        int rndGene = rnd.getRnd().nextInt(genotype.size());
5        VRPVisit v = genotype.remove(rndGene);
6        int addPoint = rnd.getRnd().nextInt(genotype.size());
7        genotype.add(addPoint,v);
8    }
```

Listing 3.4 The *Individual.mutate* method

```
1     public double evaluate() {
2         /*
3          * Build a phenotype based upon the genotype
4          * Only build the genotyoe if the phenotype has been
      set to null
5          * Return the fitness (distance)
6          */
7         if (phenotype == null) {
8             phenotype = new ArrayList<ArrayList<VRPVisit>> ();
9             ArrayList<VRPVisit> newRoute = new ArrayList<
      VRPVisit>();
10            for (VRPVisit v : genotype){
11                if (v.getDemand() + routeDemand(newRoute) >
      problem.getCapacity()){
12                    //If next visit cannot be added   due to
      capacity constraint then
13                    //start new route.
14                    phenotype.add(newRoute);
15                    newRoute = new ArrayList<VRPVisit>();
16                }
17                newRoute.add(v);
18            }
19            phenotype.add(newRoute);
20        }
21        return problem.getDistance(phenotype);
22    }
```

Listing 3.5 The *Individual.evaluate()* method

3.6   Results

ACE Doughnuts test the algorithm on the Augerat (2014a, b, c) problem instances used previously (see Tables 2.1, 2.2 and 2.3). Due to the stochastic nature of the EA, there is a significant chance that it will produce a different result each time it is run. It may be the case that the EA will evolve the same solution on successive runs, but this cannot be guaranteed. To mitigate this aspect, the algorithm is executed 20 times on each dataset.

The results in Tables 3.1, 3.2 and 2.3 show the results obtained with the EA. The results are curious, and they show that only in 3 instances (from set P) does the EA find a solution with a shorter distance than the solution found using branch and cut (Augerat 1995). In most cases, the EA can find a solution with the same number of vehicles as Augerat. Where the differences are comparatively small, some of the differences may be explained by rounding issues between the two software platforms used (Table 3.3).

Table 3.4 combines the results of the Clarke–Wright solver and VRPea and compares them to the original branch and cut results. We note that ACE Doughnuts now face the problem that confronts many organisations examining optimisation problems across a wide range of instances, and no one method of solving the problem consistently finds the desired solution. There are a number of approaches to tackle this problem as follows:

Table 3.1 EA results on the problems in set A

Problem	n	Branch and cut result			EA—Best of 20 runs			Improvement		% Dist over best
		Best or Optimal	Distance	Vehicles	Distance	Vehicles	EA Execution time	Distance	Vehicles	
A-n32-k5	31	Optimal	784	5	797.871806	5	8536.8	No	Same	1.77%
A-n33-k5	32	Optimal	661	5	662.1100976	5	7967.4	No	Same	0.17%
A-n33-k6	32	Optimal	742	6	742.6932624	6	10708.7	No	Same	0.09%
A-n34-k5	33	Optimal	778	5	786.4370017	5	10946.8	No	Same	1.08%
A-n36-k5	35	Optimal	799	5	810.0708134	5	14191.5	No	Same	1.39%
A-n37-k5	36	Optimal	669	5	682.3334967	5	12355.1	No	Same	1.99%
A-n37-k6	36	Optimal	949	6	967.5585312	6	10515.8	No	Same	1.96%
A-n38-k5	37	Optimal	730	5	736.5379862	5	11240	No	Same	0.90%
A-n39-k5	38	Optimal	822	5	845.1242148	5	10609.4	No	Same	2.81%
A-n39-k6	38	Optimal	831	6	833.2046054	6	11498	No	Same	0.27%
A-n44-k7	43	Optimal	937	6	978.6067516	6	12306.6	No	Same	4.44%
A-n45-k6	44	Optimal	944	6	963.3671093	6	11894.3	No	Same	2.05%
A-n45-k7	44	Best	1146	7	1203.253475	7	14634	No	Same	5.00%
A-n46-k7	45	Optimal	914	7	994.3830388	7	17833.6	No	Same	8.79%
A-n48-k7	48	Best	1073	7	1142.407635	7	11572.6	No	Same	6.47%
A-n53-k7	53	Optimal	1010	7	1080.655136	7	10347.7	No	Same	7.00%
A-n54-k7	54	Best	1167	7	1267.558852	7	10404.5	No	Same	8.62%
A-n55-k9	55	Optimal	1073	9	1137.547862	9	11169.6	No	Same	6.02%
A-n60-k9	60	Best	1408	9	1498.937214	9	12919.4	No	Same	6.46%
A-n61-k9	61	Best	1035	9	1145.562551	10	13153	No	No	10.68%
A-n62-k8	62	Best	1290	8	1433.095877	8	12928.6	No	Same	11.09%
A-n63-k9	63	Best	1634	9	1803.218049	9	13595.2	No	Same	10.36%
A-n63-k10	63	Best	1315	10	1437.129905	10	14008.5	No	Same	9.29%
A-n64-k9	64	Best	1402	9	1512.201803	9	13215.9	No	Same	7.86%
A-n65-k9	65	Best	1177	9	1325.965881	9	14295.7	No	Same	12.66%
A-n69-k9	69	Best	1168	9	1302.333451	9	16243.7	No	Same	11.50%
A-n80-k10	80	Best	1764	10	2081.602397	10	20912.5	No	Same	18.00%

Table 3.2 EA results on the problems in set B

Problem	n	Branch and cut result			EA—Best of 20 runs			Improvement		% Dist over best
		Best or Optimal	Distance	Vehicles	Distance	Vehicles	EA Execution time	Distance	Vehicles	
B-n31-k5	31	Optimal	672	5	676.0884123	5	5223.2	No	Same	0.61%
B-n34-k5	34	Optimal	788	5	790.8754768	5	5853	No	Same	0.36%
B-n35-k5	35	Optimal	955	5	958.8934983	5	5935.4	No	Same	0.41%
B-n38-k6	38	Optimal	805	6	809.6800685	6	6384.1	No	Same	0.58%
B-n39-k5	39	Optimal	549	5	564.6798484	5	6302.4	No	Same	2.86%
B-n41-k6	41	Optimal	829	6	839.7267182	6	6869.3	No	Same	1.29%
B-n43-k6	43	Optimal	742	6	758.1167945	6	7381	No	Same	2.17%
B-n44-k7	44	Optimal	909	7	947.3659869	7	7702.6	No	Same	4.22%
B-n45-k5	45	Optimal	751	5	754.4387545	5	7797	No	Same	0.46%
B-n45-k6	45	Optimal	678	6	702.10599	6	8902.2	No	Same	3.56%
B-n50-k7	50	Optimal	741	7	753.9575654	7	9758.9	No	Same	1.75%
B-n50-k8	50	Best	1313	8	1342.04352	8	10444.8	No	Same	2.21%
B-n51-k7	51	Optimal	1032	7	1050.017403	7	9506.9	No	Same	1.75%
B-n52-k7	52	Optimal	747	7	778.1641413	7	9502.7	No	Same	4.17%
B-n56-k7	56	Optimal	707	7	725.1954333	7	10692.9	No	Same	2.57%
B-n57-k7	57	Optimal	1153	7	1166.968036	7	10840.4	No	Same	1.21%
B-n57-k9	57	Best	1598	9	1668.647567	9	11346.8	No	Same	4.42%
B-n63-k10	63	Best	1537	10	1640.803456	10	12812.2	No	Same	6.75%
B-n64-k9	64	Optimal	861	9	930.0137125	9	13322.6	No	Same	8.02%
B-n66-k9	66	Best	1374	9	1379.908886	9	13428.9	No	Same	0.43%
B-n67-k10	67	Best	1033	10	1128.343982	10	14016.6	No	Same	9.23%
B-n68-k9	68	Best	1304	9	1402.356261	9	14285.7	No	Same	7.54%
B-n78-k10	78	Best	1266	10	1396.285482	10	18230.7	No	Same	10.29%

Table 3.3 EA results on the problems in set P

Problem	n	Branch and cut result			EA—Best of 20 runs			Improvement		% Dist over best
		Best or Optimal	Distance	Vehicles	Distance	Vehicles	EA Execution time	Distance	Vehicles	
P-n16-k8	16	Best	435	8	451.3350802	8	4103.9	No	Same	3.76%
P-n19-k2	19	Best	212	2	212.6569042	2	2753	No	Same	0.31%
P-n20-k2	20	Best	220	2	217.4155833	2	2886.5	Yes	Same	−1.17%
P-n21-k2	21	Best	211	2	212.7115397	2	3016	No	Same	0.81%
P-n22-k2	22	Best	216	2	217.8521507	2	3163.8	No	Same	0.86%
P-n22-k8	22	Best	603	8	588.7935243	9	4883.6	Yes	No	−2.36%
P-n23-k8	23	Best	554	8	531.1738035	8	4859.5	Yes	Same	−4.12%
P-n40-k5	40	Best	458	5	464.6998427	5	6630	No	Same	1.46%
P-n45-k5	45	Best	510	5	535.7071537	5	7555.3	No	Same	5.04%
P-n50-k7	50	Best	554	7	585.4231813	7	9096.2	No	Same	5.67%
P-n50-k8	50	Best	649	8	668.0002659	9	9368.2	No	No	2.93%
P-n50-k10	50	Best	696	10	730.7135563	10	9898	No	Same	4.99%
P-n51-k10	51	Best	745	10	795.5497775	10	10230.8	No	Same	6.79%
P-n55-k7	55	Best	524	7	586.1756661	7	10452.5	No	Same	11.87%
P-n55-k8	55	Best	576	8	613.2302757	7	10266.7	No	Same	6.46%
P-n55-k10	55	Best	669	10	722.2078163	10	10959.7	No	Same	7.95%
P-n55-k15	55	Best	856	15	996.3685655	16	12561.5	No	No	16.40%
P-n60-k10	60	Best	706	10	806.1750022	10	12055.7	No	Same	14.19%
P-n60-k15	60	Best	905	15	1048.562198	15	13570.7	No	Same	15.86%
P-n65-k10	65	Best	792	10	870.3763879	10	13344.3	No	Same	9.90%
P-n70-k10	70	Best	834	10	942.406153	10	15090.9	No	Same	13.00%
P-n76-k4	76	Best	589	4	683.7294245	4	17093.3	No	Same	16.08%
P-n76-k5	76	Best	631	5	730.7547836	5	16592.3	No	Same	15.81%
P-n101-k4	101	Optimal	681	4	929.9426764	4	25015.4	No	Same	36.56%

Table 3.4 A summary comparing performance of Clarke–Wright and VRPea. The best solver column denotes whether the shortest solution was found by Clarke–Wright (CW) or VRPea (EA). The difference between the best solution found by CW or EA and the benchmark branch and cut solution (Augerat 1995) is also given (The three negative values note the three situations where VRPea found a shorter result than branch and cut.)

Problem	Best solver	Increase on B&C (%)	Problem	Best solver	Increase on B&C (%)	Problem	Best Solver	Increase on B&C (%)
A-n32-k5	EA	1.77	B-n31-k5	EA	0.61	P-n16-k8	EA	3.76
A-n33-k5	EA	0.17	B-n34-k5	EA	0.36	P-n19-k2	EA	0.31
A-n33-k6	EA	0.09	B-n35-k5	EA	0.41	P-n20-k2	EA	−1.17
A-n34-k5	EA	1.08	B-n38-k6	EA	0.58	P-n21-k2	EA	0.81
A-n36-k5	EA	1.39	B-n39-k5	EA	2.86	P-n22-k2	EA	0.86
A-n37-k5	EA	1.99	B-n41-k6	EA	1.29	P-n22-k8	EA	−2.36
A-n37-k6	EA	1.96	B-n43-k6	EA	2.17	P-n23-k8	EA	−4.12
A-n38-k5	EA	0.90	B-n44-k7	CW	3.16	P-n40-k5	EA	1.46
A-n39-k5	EA	2.81	B-n45-k5	EA	0.46	P-n45-k5	EA	5.04
A-n39-k6	EA	0.27	B-n45-k6	EA	3.56	P-n50-k7	EA	5.67
A-n44-k7	CW	4.17	B-n50-k7	CW	1.16	P-n50-k8	CW	2.56
A-n45-k6	EA	2.05	B-n50-k8	EA	2.21	P-n50-k10	EA	4.99
A-n45-k7	CW	4.71	B-n51-k7	EA	1.75	P-n51-k10	CW	3.95
A-n46-k7	CW	2.82	B-n52-k7	CW	2.83	P-n55-k7	EA	11.87
A-n48-k7	CW	3.71	B-n56-k7	EA	2.57	P-n55-k8	EA	6.46
A-n53-k7	EA	7.00	B-n57-k7	EA	1.21	P-n55-k10	EA	7.95
A-n54-k7	CW	3.19	B-n57-k9	CW	3.61	P-n55-k15	CW	14.96
A-n55-k9	CW	3.50	B-n63-k10	CW	4.18	P-n60-k10	CW	12.79
A-n60-k9	CW	−0.58	B-n64-k9	CW	7.11	P-n60-k15	CW	12.37
A-n61-k9	CW	7.08	B-n66-k9	EA	0.43	P-n65-k10	CW	7.53
A-n62-k8	CW	4.87	B-n67-k10	CW	6.65	P-n70-k10	CW	7.17
A-n63-k9	CW	3.30	B-n68-k9	CW	1.10	P-n76-k4	CW	13.44
A-n63-k10	CW	4.78	B-n78-k10	CW	0.47	P-n76-k5	CW	11.93
A-n64-k9	CW	4.77				P-n101-k4	CW	12.75
A-n65-k9	CW	5.81						
A-n69-k9	CW	4.65						
A-n80-k10	CW	6.29						

1. Continue searching through optimisation methods to see if one can be found or configured to consistently produce the best results.
2. Use the method that appears most reliable and accept that some results will be sub-optimal.
3. Solve using multiple algorithms and use the best result found.

Approach 1 may yield a result given time, but there's no guarantee that a method will be found that equals or outperforms branch and cut. Approach 2 is probably the least desirable, but if it is not feasible to use cloud resources or to wait for the algorithms to execute one after the other, then it may be the only option. Approach 3 was used to produce the results in Table 3.4. Given our object-oriented approach, we can easily modify our approach in order to encompass multiple runs of the EA and the use of the Clarke–Wright algorithm. Algorithm 10 shows how such multiple runs may be undertaken. Listing 3.6 gives a Java implementation of a hybrid solver. The Clarke–Wright solver is used to construct an initial solution (line 10), and multiple runs of the EA are then made (lines 18–28). If the EA finds a shorter solution, then it replaces the best solution found so far. This implementation could be modified to execute the algorithms in parallel by having lines 10 and 21 launch the solvers on separate threads. The detail of a parallel implementation is beyond the scope of this book. If ACE Doughnuts should implement any other algorithms that extend VRPSolver, then they may be easily incorporated into this architecture.

Algorithm 10 Combining Multiple Runs

1: **Procedure** $MultipleRuns()$
2: $bestSolutionSoFar = ClarkeWright.solve()$
3: $EAruns = 20$
4: **while** $EAruns > 0$ **do**
5:    $currentSolution = EA.solve()$
6:    **if** $currentSolution.fitness < bestSolutionSoFar.fitness$ **then**
7:       $bestSolutionSoFar = currentSolution$
8:    $EAruns - -$
9: **Return** $bestSolutionSoFar$
10: **EndProcedure**

3.7 Conclusions

This chapter has introduced the Capacitated Vehicle Routing Problem which introduces a number of challenges. Like many of the problems that we will examine in this book, it requires the construction of *network* of routes. CVRP, as formulated here, is still a single-objective problem based on minimising distance.

Evolutionary Algorithms are a nature inspired technique (based on concepts from Darwinian Evolution) which have been applied to many problems in different

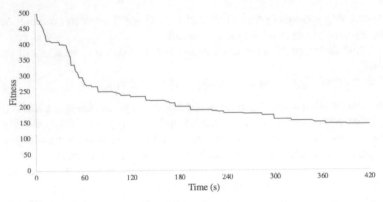

(a) A typical EA run, showing how the fitness of the best individual in the population improves over time.

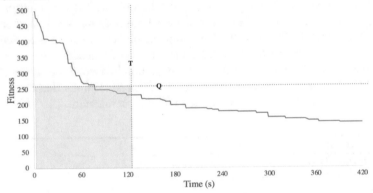

(b) The EA run shown in figure 3.3a, but the the addition of the quality threshold T and patience threshold P.

Fig. 3.3 The upper figure shows a typical EA run, plotting the fitness of the most fit individual. Over time, the fitness reduces, and the biggest improvement in fitness is seen at the beginning of the run. The lower plot shows the same run, but with the T and Q thresholds added. In this case, we need to have the fitness drop beneath Q prior to reaching T. The highlighted section of the graph denotes the zone that contains acceptable solutions

domains. One of the shortcomings of EAa is their sensitivity to the values of parameters such as population size.

Figure 3.3a plots the typical output from the EA described in this chapter showing the improvement in the fitness of the best member of the population over time. We note that the EA rapidly improves at the start, but there emerges a solution to which it can find no improvement, and the EA then *converges* upon that solution. When considering a real-world application, we can envisage two subjective variables which will affect the ability of our algorithm to satisfy the user:

- Q—the quality threshold beyond which the user will accept a solution.
- T—the time that they are willing to wait for a solution.

An example of how Q and T may relate to the run time of the EA is shown in Fig. 3.3b. An organisation such as ACE Doughnuts will not necessarily require the optimum solution, but they will require what they term to be a good solution, i.e. one that is adequate for their needs. The time limit T is subjective, a planner who is investigating a theoretical problem, perhaps the effects of adding new customers or altering the location of the depot may be willing to wait several hours for a solution, whilst in an operational scenario with customers urgently awaiting orders, a solution may be required in minutes. If we examine Fig. 3.3b, we see that Q and T divide the graph into 4 quadrants. If we find a solution in the lower left quadrant (as we do here), then we find an *acceptable solution within an acceptable time*.

```
1   private static void run(String probName) {
2       /*
3        * Solve the instance named in  <probName>
4        */
5       CVRPProblem myVRP = VRPProblemFactory.buildProblem(
    probName);//Load instance from file
6
7       System.out.print(probName + ",");
8
9       //Solve using Clarke & Wright
10      myVRP.solve(new ClarkeWright());
11      ArrayList<ArrayList<VRPVisit>> bestSolution = myVRP.
    getCVRPSolution();
12      double bestDist = myVRP.getDistance();
13      System.out.println("Clarke-Wright distance =" +
    bestDist);
14
15      //Solve using the Evolutionary Algorithm
16      //As the Evolutionary Algorithm is stochastic, we
    repeat 20 times and report the best and average results
17
18      for (int count = 0; count < 20; count ++){
19          RandomSingleton rnd = RandomSingleton.getInstance()
    ;
20          rnd.setSeed(count);
21          myVRP.solve(new VRPea());
22          System.out.println("EA distance =" + myVRP.
    getDistance());
23          if (myVRP.getDistance() < bestDist){
24              //if we have found a better solution
25              bestDist = myVRP.getDistance();
26              bestSolution = myVRP.getCVRPSolution();
27          }
28      }
29      //Print out final solution
30      System.out.println("Solution");
31      System.out.println("Total distance = " + bestDist);
32      for(ArrayList<VRPVisit> route : bestSolution){
33          for (VRPVisit v : route){
34              System.out.print(v+":");
35          }
36          System.out.println();
37      }
38  }
39 }
```

Listing 3.6 The hybrid solver

References

Augerat, P. 1995. Approche polyèdrale du problème de tournées de véhicules. (Polyhedral approach of the vehicle routing problem). PhD Thesis, Grenoble Institute of Technology, France. https://tel.archives-ouvertes.fr/tel-00005026.

Augerat, P. 2014a. VRP-REP: Augerat 1995 Set A. http://www.vrp-rep.org/datasets/item/2014-0000.html.

Augerat, P. 2014b. VRP-REP: Augerat 1995 Set B. http://www.vrp-rep.org/datasets/item/2014-0001.html.

Augerat, P. 2014c. VRP-REP: Augerat 1995 Set P. http://www.vrp-rep.org/datasets/item/2014-0009.html.

Colorni, A., M. Dorigo, and V. Maniezzo. 1992. An Investigation of Some Properties of an "Ant Algorithm". In *Parallel Problem Solving from Nature 2, PPSN-II, Brussels, Belgium, September 28–30, 1992*, ed. R. Männer and B. Manderick, 515–526. Amsterdam: Elsevier.

Dantzig, G.B., and D.R. Fulkerson. 1954. Minimizing the Number of Tankers to Meet a Fixed Schedule. *Naval Research Logistics Quarterly* 1 (3): 217–222. https://onlinelibrary.wiley.com/doi/pdf/10.1002/nav.3800010309. https://onlinelibrary.wiley.com/doi/abs/10.1002/nav.3800010309.

Dantzig, G.B., and J.H. Ramser. 1959. The Truck Dispatching Problem. *Management Science* 6 (1): 80–91. https://doi.org/10.1287/mnsc.6.1.80.

Dantzig, G., R. Fulkerson, and S. Johnson. 1954. Solution of a Large-Scale Traveling-Salesman Problem. *Journal of the Operations Research Society of America* 2 (4): 393–410. Publisher: INFORMS. http://www.jstor.org/stable/166695.

Darwin, C. 1859. *On the Origin of Species by Means of Natural Selection*. London: John Murray.

Eiben, A.E., and S.K. Smit. 2012. Evolutionary Algorithm Parameters and Methods to Tune Them. In *Autonomous Search*, ed. Y. Hamadi, E. Monfroy, and F. Saubion, 15–36. Heidelberg: Springer.

Goldberg, David. 1989. *Genetic Algorithms in Search, Optimization and Machine Learning*. Boston: Addison-Wesley Longman Publishing Co., Inc.

Hertz, J., A. Krough, and R.G. Palmer. 1991. *Introduction to the Theory of Neural Computation*, vol. 44. Physics Today.

Holland, J.H. 1975. *Adaptation in Natural and Artificial Systems*. Ann Arbor: University of Michigan Press.

Kephart, J.O. 1994. A Biologically Inspired Immune System for Computers. In *Artificial Life IV: Proceedings of the Fourth International Workshop on the Synthesis and Simulation of Living Systems*, 130–139. MIT Press.

Chapter 4
Solving Problems That Have Dual Solution Characteristics

Abstract Many real-world delivery problems must take into account multiple characteristics when searching for the desired solution (e.g. costs, delivery time or environmental impact). The techniques for finding solutions to the Vehicle Routing Problem discussed previously (e.g. Clarke and Wright Chap. 2 and Evolutionary Algorithms Chap. 3), two solution characteristics became apparent; the number of vehicles used and the distance travelled. The approaches discussed previously searched for solutions based on the minimising of a single objective value. This single objective value could be distance, the quantity of vehicles or some utility value based on a combination of both. For many real-world problems, a solution that focuses on one characteristic at the expense of the other may not be appropriate, nor may be desirable to present a solution based on a combination of the two—which may produce a solution that does not satisfactorily address either characteristic. To address this problem, this chapter introduces a theme that recurs through this book; incorporating user choice into the problem-solving process. We encourage user choice by presenting the user with a structured set of solutions rather than one solution created by the optimiser.

4.1 Dealing with Solutions with Twin-Characteristics

Within our formulation of the ACE Doughnuts problem Sect. 2.2, we optimised based upon a single objective (distance) despite there being two solution characteristics (vehicles and distance). Rather than concentrating on objectives, we choose to focus upon solution characteristics, seeking to generate multiple solutions with a range of characteristics which may then be presented to the user allowing them to make the final, informed, choice of solution.

Before we discuss solution characteristics, it is useful to discuss how we handle dual objectives within optimisation. Let us assume that our objectives are x and y and that in both cases we seek to minimise them. Both of the objectives are solution characteristics, altering the solution will alter x and y. As they are both characteristics then we can assume that they are in some way related, altering the solution to change x will usually result in a change in y and vice-versa. Let us take a practical example,

N. Urquhart, *Nature Inspired Optimisation for Delivery Problems*,
Natural Computing Series, https://doi.org/10.1007/978-3-030-98108-2_4

suppose we have a vehicle routing problem with the objectives emissions e (CO_2) and average delivery time t (the average time taken to reach each customer). We have the option of making use of diesel powered vans or cargo bikes. Diesel powered vans are quick, but produce CO_2 whilst cargo bikes are far slower but produce no emissions. If we wish to decrease t we use more vans as they are quicker, but this increases e, if we use more cargo bikes then we reduce e, but increase t. We can describe e and t as being *conflicting*. When two characteristics are conflicting then an attempt to reduce one generally results in an increase in the other. Conversely, if we examine emissions e and running costs rc then we might find that these objectives are *non-conflicting* reducing emissions by making more use of cargo bikes, also reduces running costs as the bikes are cheaper to operate.

There would appear to be three choices of solution available to us:

1. Find a solution that optimises e at the expense of t;
2. Find a solution that optimises t at the expense of e;
3. Find a solution that optimises t and e such that neither is optimal, but both are acceptable to the user.

The third type of solution is often referred to as a *trade-off*. The decision as to whether to give priority to e or t is not really a decision for us to make, ultimately that decision lies in the hands of the domain expert. The issue could be summarised as *to what degree is saving the environment more important than making deliveries quicker?* This is not necessarily a question that can be answered by an algorithm, but one where the domain expert may be best placed to make a final choice. The factors taken into account could include organisational or political policies, legislation, economics or public relations considerations. In this situation, presenting the user with a set of representative solutions that demonstrate the range of choices available becomes an attractive option. The range of solutions presented should ideally include the solutions that demonstrate the maximum values of each characteristic (solution types 1 and 2 above) and a number of solutions that illustrate possible trade-offs, which showcase existence of compromise solutions. A useful set of trade-offs would include solutions that illustrate a range of values for e with corresponding values for t.

The set of solutions presented to the user must not be too small, which limits user choice, and not too large. If the set is too large, we run the risk of the user being overwhelmed with choice and unable to make an effective decision. It is no longer the role of the optimiser to find the optimal solution and present it to the user. It is now the role of the optimiser to provide a set of solutions that demonstrate a range of solution characteristics which showcase the options available to the user and *support* them in making the final choice of solution.

4.2 Pareto Dominance

In order to effectively manage twin-characteristic solutions, we need a means of ranking two solutions and establishing which is the more fit. For a single characteristic, this becomes trivial; we simply compare the numerical fitness values. One means of dealing with bi-objective problems is to combine the objectives into a single fitness value f thus:

$$f = (xa) + (yb)$$

where a and b are weights applied to the objectives. This has the immediate drawback that values for a and b must be set, which involves giving one a higher weighting, and therefore, implied importance over the other. It can be argued that by setting such weights in advance we are directing the optimisation towards a specific part of the search space.

Rather than take the approach of combining solution characteristics into a single fitness value, we can use the technique of *Pareto Dominance* to compare solutions. The Italian economist and engineer Vilfredo Pareto (2021) established the concept of *Pareto Dominance*. ,, allowing us to compare two solutions and examine if one *dominates* the other. Assuming that we have two solutions s_x and s_y:

s_x is said to *dominate* s_y if

- s_x is no worse than s_y in all characteristics;
- s_x is better than s_y in at least one characteristic.

We can use the concept of dominance to rank solutions when undertaking selection within an EA, allowing us to take into account both characteristics and avoiding the need to combine them into a fitness value.

To illustrate our point we can plot solutions on a 2D plot, (Fig. 4.1a) which shows a population of solutions plotted against 2 characteristics. Both characteristics are in the range 1–20 and need to be minimised. In an ideal world, we would like a solution that minimises both characteristics (1, 1), but such an ideal solution is unlikely to exist if the characteristics are opposed to any degree. The population includes two solutions (20, 1) and (1, 20) which optimise one of the characteristics at the expense of the other. It is worth noting that for many optimisation problems, the number of feasible solutions in the solution space may run into thousands, this example has relatively few in order to improve clarity.

In Fig. 4.1b, we take a solution at random A and apply the Pareto dominance rules discussed above. Every solution in the shaded area is dominated by A. When selecting a solution, there is no logical reason for the user to select any of the three dominated solutions in preference to A. In each of the two objectives A is at least equal, and in at least one (if not both) is better.

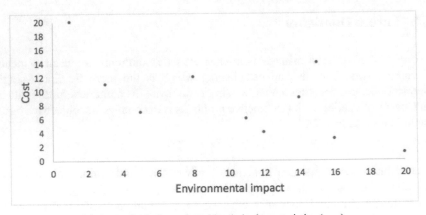

(a) A set of solutions plotted by their characteristics (c, e).

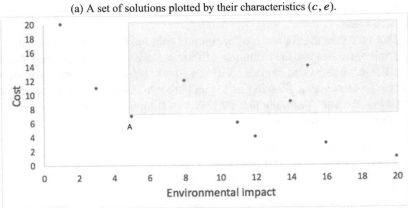

(b) The area of the solution space that is *dominated* by solution A

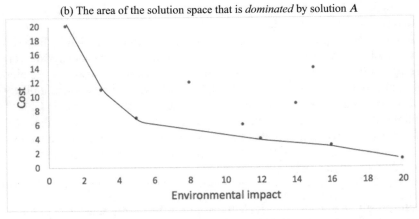

(c) The the set of non-dominated solutions highlighted as a front.

Fig. 4.1 The basic principles of Pareto dominance and the identification of a non-dominated front

	c	e	
	~~12~~	~~8~~	Dominated by A
	9	~~14~~	Dominated by A
	~~14~~	~~15~~	Dominated by A
A	5	7	

We can see from the above table, that if the user is solely concerned with characteristics e and c then they will have no interest in any of the three solutions that have been dominated. If we examine Fig. 4.1b closely we note that A is not dominated by any other solution, therefore we term it *non-dominated*. The set of solutions that we wish to present to the user is a *non-dominated front*, that is to say the set of non-dominated solutions. Figure 4.1c shows the set of non-dominated solutions presented as a front.

The set of non-dominated solutions is

c	e	
20	1	minimises e
11	3	trade-off
7	5	trade-off
5	12	trade-off
3	16	trade-off
1	20	minimises c

This non-dominated set presents the choices available to the user, at the head and tail of the list we can see the extreme solutions that minimise c and e at the expense of the opposite objective. We note that the second solution is an interesting trade-off; a large reduction in c can be obtained for a (relatively) small increase in e. In this way, the user can easily view the extent to which c and e may be traded against one another.

Algorithm 11 shows the manner in which the non-dominated front may be extracted from a population. Each member of the population is considered in turn (line 3) and is checked against every other member of the population (line 5) to see if any of them dominated it (line 6).

Algorithm 11 Finding a non dominated front.

1: **Procedure** FindNonDomFront(population[])
2: *nonDom* = []
3: **for** Solution *possible* **in** *population* **do**
4:     *dominated* = **False**
5:     **for Solution** *other* **in** *population* **do**
6:         **if** *other.dominates(possible)* **then**
7:             *dominated* = **True**
8:             **break**
9:     **if** !*dominated* **then**
10:         *nonDom.add(possible)*
11: **Return** *nonDom*
12: **EndProcedure**

4.3 An Evolutionary Algorithm to Find Non-dominated Fronts

As a population-based heuristic, the Evolutionary Algorithm is easily adaptable to finding Pareto fronts. The evolutionary mechanism continues to be used to evolve a population, when the algorithm has completed a non-dominated front can be extracted. In our previous EA (see Sect. 3.3) our fitness function was a single objective (distance) which had to be minimised, this made selection and replacement straightforward. In our modified version of the EA, we will carry out selection based on dominance, rather than fitness.

Algorithm 12 Selection based on domination.

1: **Procedure** $DominatedTournament(population[])$
2: $x = population.random()$
3: $y = population.random()$
4: **if** $x.dominates(y)$ **then**
5:     **return** x
6:     **else**
7:     **return** y
8: **EndProcedure**

When using the Evolutionary Algorithm it is important to understand that the front returned is a *non-dominated* front, but not necessarily the *Pareto optimal* front. The Pareto optimal front represents the optimal set of non-dominated solutions taking into account the *entire* solution space. As we are dealing with NP-Complete problems (see Sect. 1.1), we can no more guarantee to find the Pareto Optimal set within a reasonable time than we can guarantee to find the optimal solution. Because of these properties, we use the EA to find a non-dominated front of high-quality solutions rather than the specific Pareto Optimal front.

4.4 Case Study: Routes Versus Customer Service

We consider, once more, the logistics problems of Ace Doughnuts (see Sect. 2.2), previously ACE Doughnuts tried a range of techniques to produce solutions to their problems. These techniques resulted in the production of a single solution. Increasingly the logistics manager finds this approach troublesome as the solution presented does not always meet their needs.

An analysis suggests that there are two characteristics which are important to ACE Doughnuts when considering a solution:

- **Number of routes**: The number of vehicles and drivers required.
- **Customer service**: A metric based on the time taken for each customer to receive their order, each customer wants their order as quickly as possible.

It is useful to consider these characteristics in more detail, in particular the *range* of each characteristic. When considering routes, the range will be defined by

- The *upper bound* will be n (where n is the number of customers) representing a solution with a dedicated route for each customer; this should provide the best customer service by having the deliveries made in the shortest time.
- The *lower bound* will be the least number of routes required for each problem instance. In these examples, we know this value has been established previously Augerat (1995) and is given as k in the problem name.

ACE Doughnuts would like to be able to select a solution based on varying values of the above objectives. In practice, they may have differing numbers of vehicles available each day due to staff absences or mechanical troubles. There are some days when they may opt to run fewer vehicles if this can be achieved without significant effects on customer service.

A Revised Representation

Prior to optimising the problem as a bi-characteristic problem, ACE Doughnuts modify their existing algorithm (see Sect. 2.2) to be able to optimise either characteristic (routes or customer service) as a single objective. Table 4.1 shows a sample of the initial results. Given that we know the upper and lower bounds of the Routes characteristic, we become aware that the results obtained are poor, when using the Customer Service characteristic, we would expect the number of routes to increase to n for the one route per customer solution.

The problem faced by ACE Doughnuts is that the representation and decoder used (see Sect. 2.2) is biased towards reducing the number of routes. The decoder allocates customers to a route until the available capacity is used up. It is not possible to have a route that has large amounts of spare capacity.

ACE Doughnuts need a representation that accommodates solutions which minimise customer service at the expense of routes, rather than one that always minimises route. To achieve this, we take the previous representation and modify it so that each "gene" also contains a Boolean flag to distinguish whether this delivery should be made at the start of a new route.

Table 4.1 A sample of the initial (unsatisfactory) results achieved with the existing algorithm using no of routes and customer service as single objectives. For this extract, the best run from 10 is presented

Problem	Optimisation Criterion			
	Routes		Customer service	
A-n32-k5	5	7203.50	5	5466.66
A-n33-k5	5	5094.28	6	3428.42
A-n33-k6	6	3976.43	7	2887.23
A-n34-k5	5	6110.60	6	4360.70
A-n36-k5	5	8597.31	6	5498.88
A-n37-k5	5	6051.20	5	4443.99
A-n37-k6	6	6966.94	8	4429.09
A-n38-k5	5	6443.86	6	4507.82
A-n39-k5	5	8542.63	6	5047.64
A-n39-k6	6	6426.50	7	4501.60
A-n44-k7	6	9027.99	7	5960.35
A-n45-k6	6	8798.21	7	6258.67
A-n45-k7	7	9335.07	8	6751.62
A-n46-k7	7	7868.38	7	5872.94
...	...	...	...	...

Assuming that we have customer deliveries A to E with demands as follows:

Cust	Demand
A	2
B	3
C	2
D	1
E	2

An example of a valid genotype would be

A,False	B,True	C,False	D,True	E,True

Note that the "gene" now comprises <customerID>, <New route>. Assuming that we have a capacity of 5, the phenotype decodes to the following genotype under the current decoder:

Route	Customers
1	A B
2	C D E

Using the modified decoder (taking into account the New Route flags), the following phenotype would be created:

Route	Customers
1	A
2	B C
3	D
4	E

The sequence of "genes" still determines the order in which they are added to the network. The decoder is now modified (see Listing 4.2) so that in any case where the new route flag is 'true' (line 11) a new route is started (line 19), regardless of remaining capacity within the current route.

```
1    private class Gene {
2        /*
3         * Represents a single Gene,
4         *    visit - the visit being made
5         *    newRoute - true if this visit should be added to
     the start of a new route
6         */
7        private boolean newRoute = false;
8        private VRPVisit visit = null;
9
10       public Gene(VRPVisit aVisit){
11           this.visit = aVisit;
12           this.newRoute = false;
13       }
14
15       public boolean newRoute(){
16           return this.newRoute;
17       }
18
19       public VRPVisit visit(){
20           return this.visit;
21       }
22
23       public void setNewRoute(boolean val){
24           this.newRoute = val;
25       }
26
27       public void flipNewRoute(){
28           this.newRoute = ! this.newRoute;
29       }
30
31       public String toString(){
32           return this.visit.toString() + "New Route "+ this.
     newRoute;
33       }
34
35   }
```

Listing 4.1 The Gene class (within BiObjectiveIndividual.java)

```
1     private void decode() {
2        /*
3         * Build a phenotype based upon the genotype
4         * Only build the genotype if the phenotype has been
      set to null
5         */
6        if (phenotype == null) {
7            phenotype = new ArrayList<ArrayList<VRPVisit>> ();
8            ArrayList<VRPVisit> newRoute = new ArrayList<
      VRPVisit>();
9            for (Gene g : genotype){
10               VRPVisit v = g.visit();
11               if (g.newRoute()){//Create a new route, if the
      gene specifies a new route
12                   phenotype.add(newRoute);
13                   newRoute = new ArrayList<VRPVisit>();
14               }
15               if (v.getDemand() + routeDemand(newRoute) >
      problem.getCapacity()){
16                   //If next visit cannot be added  due to
      capacity constraint then
17                   //start new route.
18                   phenotype.add(newRoute);
19                   newRoute = new ArrayList<VRPVisit>();
20               }
21               newRoute.add(v);
22           }
23           phenotype.add(newRoute);
24       }
25   }
```

Listing 4.2 The revised decode() method

A genotype with all of the new route flags set to false will, in practice, behave as previously with the number of routes being minimised, whilst a genotype with the flags all set to true will result in a one route per customer solution. An advantage of having the new route flag grouped with the customer is that the crossover operator does not need to be modified, any customer copied to a child retains the preference for a new route from the parent. The mutation operator has to be modified such that there is a 0.5 chance of the previous (swap) mutation being selected versus the random selection and inversion of a new-route bit.

One potential drawback of this approach is the increase in the search space size. Previously when using a permutation the search space was $n!$, in each solution there is now an n-length binary string as well. This means that for each possible permutation there are 2^n possible strings associated with it.

This new representation increases the risk that it may not be possible to find solutions in a reasonable time. Consider the following table which shows the increased in the search space size for small values of n:

n	n!	$n!n^2$
1	1	2
2	2	8
3	6	48
4	24	384
5	120	3840

Tables 4.2, 4.3 and 4.4 show the results obtained when optimising on single objectives with the updated representation. We can see that the lower bound route is usually found when using routes as the objective and the upper bound is normally found when using customer service as the objective. We also note that the customer service metric value increases when the number of routes lessens and decreases when the number of routes increases. It should be noted that the distance values found are not as low as with the original representation.

Visualisations of the solutions found to the same problem with different objectives may be seen in Fig. 4.2a–c.

So far ACE Doughnuts have found the ends of their non-dominated fronts. In effect they are aware of the extreme solutions, what they require now are the trade-off solutions that comprise the middle sections of the front.

Methodology

ACE Doughnuts modify their previous architecture in order to implement the new algorithm. A new class to represent the Individual is required that incorporates the updated representation. A refactoring is applied to create an abstract class *EAIndividual* (see Listing 4.3); this defines the basic operations that an individual must implement in order to be part of an Evolutionary Algorithm. The individual used within the new EA is represented by the class *BiObjectiveIndividual* which extends *EAIndividual*.

As this algorithm will use domination to select individuals for parenting and replacement, an interface Domination (see Listing 4.4) has been created, which defines the methods required for an object that will use dominance to compare with other objects. We create a class NonDominatedPop (see Listing 4.5) which extends ArrayList<Domination> to manage a population of objects that implement Domination. This class also allows for the extraction of a non-dominated front from a larger set. The architecture used may be seen in Fig. 4.3.

Table 4.2 The results obtained on the A set of problem instances with number of routes and customer service as objectives. Results are presented with the problem formulated for single objective and bi-characteristic

Problem	Single objective: Routes		Single objective: Customer service		Grand front				
	Routes	Cust service	Cust service	Routes	Front size	Min routes	Max routes	Min cust service	Max cust service
A-n32-k5	5	8051.56	1870.38	32	27	5	31	1870.38	6096.56
A-n33-k5	5	5231.27	1309.34	33	29	5	33	1309.34	4264.70
A-n33-k6	6	4748.40	1272.05	33	28	6	33	1272.05	3689.39
A-n34-k5	5	6351.24	1577.57	34	30	5	34	1577.57	5474.27
A-n36-k5	5	9225.72	1948.28	36	31	5	36	1948.28	6923.66
A-n37-k5	5	6748.63	1377.85	37	33	5	37	1377.85	4997.87
A-n37-k6	6	7599.54	1906.09	37	28	6	36	1906.09	6017.81
A-n38-k5	5	7256.60	1537.66	38	34	5	38	1537.66	5852.53
A-n39-k5	5	8670.88	1809.13	39	31	5	38	1809.13	6915.72
A-n39-k6	6	8184.30	1744.64	39	32	6	39	1744.64	5902.02
A-n44-k7	6	9445.71	2084.41	44	39	6	44	2084.41	7679.98
A-n45-k6	6	9481.10	2142.22	45	39	6	45	2142.22	9148.54
A-n45-k7	7	10038.97	2557.14	45	34	7	44	2557.14	8514.19
A-n46-k7	7	8626.35	2019.97	46	37	7	46	2019.97	7003.61
A-n48-k7	7	11557.80	2718.47	48	41	7	48	2718.47	9596.71
A-n53-k7	7	12053.70	2513.26	53	45	7	53	2513.26	9941.43
A-n54-k7	7	13655.96	2887.46	54	44	7	54	2887.46	10981.54
A-n55-k9	9	8467.67	2188.87	55	45	9	55	2188.87	7038.40
A-n60-k9	9	12847.45	3164.88	60	48	9	60	3164.88	10740.12
A-n61-k9	10	8718.13	2235.44	61	51	9	61	2235.44	9004.62
A-n62-k8	8	16568.38	3553.09	62	50	8	62	3553.09	14151.37
A-n63-k9	9	18002.04	4360.97	63	48	9	63	4360.97	16770.94
A-n63-k10	10	11379.63	2880.56	63	51	10	63	2880.56	9621.44
A-n64-k9	9	16625.32	3816.04	64	51	9	64	3816.04	14691.63
A-n65-k9	9	12078.05	2861.47	65	56	9	65	2861.47	11859.38
A-n69-k9	9	13575.17	2756.89	69	55	9	68	2756.89	11351.66
A-n80-k10	10	27076.50	5576.90	80	63	10	80	5576.9	23670.4

Table 4.3 The results obtained on the B set of problem instances with number of routes and customer service as objectives. Results are presented with the problem formulated for single objective and bi-characteristic

Problem	Single objective: Routes		Single objective: Customer service		Grand front			Min cust service	Max cust service
	Routes	Cust service	Cust service	Routes	Front size	Min routes	Max routes		
B-n31-k5	5	7381.08	1761.17	31	27	5	31	1761.17	5921.31
B-n34-k5	5	10250.41	2018.42	34	30	5	34	2018.42	7398.73
B-n35-k5	5	11496.26	2555.93	35	30	5	34	2555.93	9447.58
B-n38-k6	6	7995.29	1990.25	38	33	6	38	1990.25	6501.18
B-n39-k5	5	8543.22	1782.93	39	36	5	39	1782.93	6546.12
B-n41-k6	6	8606.98	2041.54	41	28	6	41	2041.54	7090.47
B-n43-k6	6	8270.04	1866.66	43	35	6	43	1866.66	6789.20
B-n44-k7	7	8928.42	2218.45	44	36	7	44	2218.45	7321.34
B-n45-k5	5	10268.75	1856.27	45	41	5	45	1856.27	8508.67
B-n45-k6	6	7938.87	1757.80	45	40	6	45	1757.80	7248.43
B-n50-k7	7	8880.08	1935.64	50	42	7	50	1935.64	7591.79
B-n50-k8	8	11474.16	2865.77	50	45	8	50	2865.77	9617.23
B-n51-k7	7	11773.00	2721.50	51	43	7	50	2721.50	11138.39
B-n52-k7	7	10419.47	2284.10	52	42	7	52	2284.10	8444.09
B-n56-k7	7	11307.15	2213.22	56	49	7	56	2213.22	8348.30
B-n57-k7	8	16208.24	3521.09	57	57	8	57	3521.09	13293.97
B-n57-k9	9	16735.25	4401.40	57	46	9	56	4401.40	15508.44
B-n63-k10	10	16566.65	4169.62	63	58	10	63	4169.62	14176.77
B-n64-k9	9	10197.69	2268.46	64	67	9	64	2268.46	9201.58
B-n66-k9	9	16199.46	3737.35	66	56	9	66	3737.35	14593.19
B-n67-k10	10	11008.46	2619.00	67	53	10	67	2619.00	9170.91
B-n68-k9	9	18462.80	3929.41	68	55	9	68	3929.41	15728.21
B-n78-k10	10	17530.67	3564.38	78	60	10	78	3564.38	14473.82

Table 4.4 The results obtained on the P set of problem instances with number of routes and customer service as objectives. Results are presented with the problem formulated for single objective and bi-characteristic

Problem	Single objective: Routes		Single objective: Customer service		Grand front				
	Routes	Cust service	Cust service	Routes	Front size	Min routes	Max routes	Min cust service	Max cust service
P-n16-k8	8	656.97	379.92	16	8	8	15	379.92	560.69
P-n19-k2	2	2517.44	479.94	19	17	2	18	479.94	2091.10
P-n20-k2	2	2742.65	510.85	20	18	2	19	510.85	2390.61
P-n21-k2	2	3075.31	520.85	21	19	2	20	520.85	2439.17
P-n22-k2	2	3241.93	547.02	22	21	2	22	547.02	2784.23
P-n22-k8	8	1227.24	582.75	22	15	8	22	582.75	1024.37
P-n23-k8	9	1098.57	549.26	23	14	9	22	549.26	872.33
P-n40-k5	5	4478.92	930.91	40	36	5	40	930.91	3726.59
P-n45-k5	5	5826.21	1094.57	45	41	5	45	1094.57	4922.58
P-n50-k7	7	4838.40	1116.28	50	43	7	49	1116.28	4179.89
P-n50-k8	9	3935.07	1116.28	50	41	9	50	1116.28	3357.73
P-n50-k10	10	3561.68	1116.28	50	40	10	50	1116.28	3166.49
P-n51-k10	11	3713.17	1201.17	51	38	11	51	1201.17	3159.26
P-n55-k7	7	5638.62	1220.84	55	48	7	55	1220.84	4985.53
P-n55-k8	7	5544.74	1220.84	55	46	7	55	1220.84	4868.77
P-n55-k10	10	4403.96	1220.84	55	44	10	55	1220.84	3649.66
P-n55-k15	16	2885.30	1220.84	55	37	16	55	1220.84	2716.22
P-n60-k10	10	5226.04	1385.97	60	50	10	60	1385.97	4509.92
P-n60-k15	16	3522.87	1387.02	60	43	15	60	1387.02	3667.16
P-n65-k10	10	6034.74	1552.26	65	52	10	65	1552.26	5410.35
P-n70-k10	10	6904.60	1670.14	70	56	10	70	1670.14	6548.58
P-n76-k4	4	19629.87	1815.43	76	66	4	76	1815.43	16470.53
P-n76-k5	5	15418.45	1815.43	76	66	5	76	1815.43	13618.88
P-n101-k4	4	35245.80	2494.71	101	89	4	101	2494.71	30399.89

Fig. 4.2 The same problem
(P-n20-k2) optimised for
customer service, no of
routes and distance as the
objectives

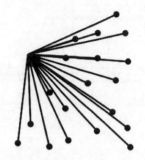

(a) objective = customer service
routes = 20
customer service = 510.85
distance = 1021.7.

(b) objective = routes
routes = 2
customer service = 2638.2
distance = 530.68.

(c) objective = distance
routes = 2
customer service = 2894.47
distance = 370.69.

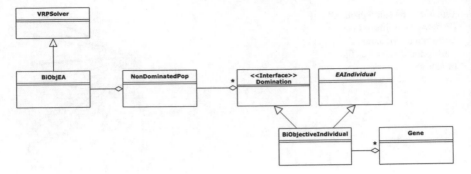

Fig. 4.3 The architecture used by ACE Doughnuts for their solver

```
1   public abstract class EAIndividual    {
2       //The phenotype is a set of routes created from the
        genotype
3       protected ArrayList<ArrayList<VRPVisit>> phenotype;
4
5       //THe problem being solved
6       protected CVRPProblem problem;
7
8       public EAIndividual(){}
9       public EAIndividual( CVRPProblem prob) {
10          /*
11           * Constructor to create a new random genotype
12           */
13          problem = prob;
14      }
15
16      public  EAIndividual (CVRPProblem prob, EAIndividual
        parent1, EAIndividual parent2){
17          /*
18           * Create a new Individual based on the recombination
        of genes from <parent1> and <parent2>
19           */
20          problem = prob;
21
22      }
23      public abstract void mutate();
24      public abstract double evaluate();
25      public abstract ArrayList getPhenotype();
26      public abstract EAIndividual copy() ;
27          //Create a new individual that is a direct copy of this
        individual
28  }
```

Listing 4.3 The EA Individual class

```
1   public interface Domination {
2       public boolean dominates(Domination other);//True if this
        object dominates the other
3       public double[] getVector();//Return the objectives as an
        array of double
4   }
```

Listing 4.4 The Domination interface

```
1   public class NonDominatedPop extends ArrayList<Domination> {
2
3       private RandomSingleton rnd = RandomSingleton.getInstance()
        ;
4       //Note that we use the RandomSingleton object to generate
        random numbers
5
6       public NonDominatedPop() {super();}//Default constructor.
7
8       public NonDominatedPop(int size, CVRPProblem theProblem) {
9           //Create a new population with <size> random solutions
        to <theProblem>
10          //Initialise population with random solutions
11          BiObjectiveIndividual best = null;
12          for (int count=0; count < size; count++){
13              BiObjectiveIndividual i = new BiObjectiveIndividual
        (theProblem);
14              if (best == null)
15                  best = i;
16              if (i.evaluate() < best.evaluate())
17                  best = i;
18              this.add(i);
19          }
20      }
21
22      public Domination getDominator(){
23          /*
24           * Select two individuals at random and if one
        dominates the other return it
25           */
26          Domination x = this.get(rnd.getRnd().nextInt(this.size
        ()));
27          Domination y = this.get(rnd.getRnd().nextInt(this.size
        ()));
28          if (x.dominates(y))
29              return x;
30          else
31              return y;
32      }
33
34      public NonDominatedPop extractNonDom(){
35          /*
36           * Return a population containing ONLY the non
        dominated individuals
37           */
38          NonDominatedPop result = new NonDominatedPop();
39          for (Domination current : this){
40              boolean dominated =  false;
41              for (Domination member : this){
42                  if (member.dominates(current))
43                      dominated = true;
44              }
45              if (! dominated)
46                  result.add(current);
47          }
48          return result;
49      }
```

Listing 4.5 The NonDOminatedPop container

The evolutionary loop may be seen in Listing 4.6. The reader will note that we have made some changes from earlier single objective EA (see Listing 3.1. Rather than creating one child per loop, we now use an inner loop (line 6) to create a batch of children (determined by the *CHILDREN* parameter). The children are created, mutated and evaluated as before and then added to the population (line 21). Having added all of the children, the population is then reduced to only the non-dominated individuals (line 21). The identification of the non-dominated individuals is a computationally intensive process, so we create a batch of children before updating the population. As our algorithm is stochastic we repeat each execution 10 times, we continue to use the same EA parameters as before.

```java
1    public void solve() {
2        population = new NonDominatedPop (INIT_POP_SIZE, super.
     theProblem);
3        evalsBudget = evalsBudget - INIT_POP_SIZE;//account for
     initial solutions in pop.
4
5        while(evalsBudget >0) {
6            for (int count =0; count < CHILDREN; count ++){
7                //Create child
8                BiObjectiveIndividual child = null;
9                if (rnd.getRnd().nextDouble() < XO_RATE){
10                   //Create a new Individual using recombination,
     randomly selecting the parents
11                   child = new BiObjectiveIndividual(super.theProblem
     , (BiObjectiveIndividual) population.getDominator(),(
     BiObjectiveIndividual) population.getDominator());
12               }
13               else{
14                   //Create a child by copying a single parent
15                   child = ((BiObjectiveIndividual) population.
     getDominator()).copy();
16               }
17               child.mutate();
18               child.evaluate();
19               evalsBudget --;
20
21               population.add(child);
22           }
23           population = population.extractNonDom();
24       }
25   }
```

Listing 4.6 The NonDominated EA main loop

A useful feature of non-dominated fronts is the ability to combine multiple runs into a single *grand front*. As we have a non-dominated front from each of the 10 runs, we can create a grand front as follows:

$$grand Front = nd(p_1 \cup p_2 \cup ...p_{10})$$

where $nd(s)$ is a function that returns a set comprising the non-dominated members of s p_n is the set of solutions returned from run n of the EA.

4.5 Results

Tables 4.2, 4.3 and 4.4 show the results obtained by ACE Doughnuts. Firstly we note that in every case the best (lowest) values obtained for each characteristic within the grand front matches that found when optimising as a single objective problem. If we examine the sizes of the grand fronts produced we note that the smallest is 8, the largest is 89, with an average choice of 41 solutions for each problem.

Figure 4.6 shows the fronts found at the start and end of a run of the EA on the largest problem instance (P-n101-K4). For clarity, the caption gives the values of the solutions at each end of the fronts. Note that the minimum number of routes is not improved, but that the optimum has been found after the initial population of 1,000 solutions has been generated. The same applics to finding the route with the minimum customer service cost, which is generated by any genotype which is all True in the new route flag within each gene.

Anew random chromosome is created (see Listing 4.7) with all of the *newRoute* flags set to *True* or to *False*. This ensures that the initial population contains a range of solutions that includes several "one route per customer" examples as well as a range of solutions which will only commence new routes based on vehicle capacity. Reference once more to Fig. 4.6 shows that the main contribution of the EA is to increase the number of *trade-off* solutions, especially those with smaller numbers of routes. However, it is these trade-off solutions that are of most interest to the user who is interested in finding out about the availability of solutions to match very specific objectives.

The approach taken by ACE Doughnuts combines the results of 10 runs into a single *grand front*. It is useful to know if the grand front represents any significant improvement. If we select the B-n78-k10 problem instance (at random), we can plot

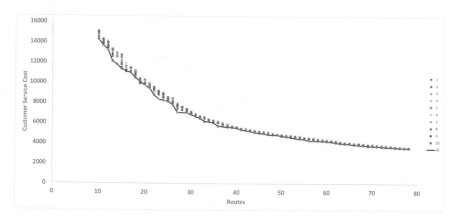

Fig. 4.4 All 10 fronts and the resulting grand front (shown as a line) for the B-n78-k10 problem instance

each of the 10 individual fronts and the grand front created from them Fig. 4.4. It can be seen from the plot that the grand front (represented by the line) consists of points (represented by different shapes) from a variety of runs.

4.6   The Hypervolume Indicator

Directly comparing Pareto fronts can be problematic, we need to compare two sets of solutions each of which has a set of characteristics. We could compare performance across individual characteristics, for example, averaging values, but this only takes into account one characteristic and an average will not necessarily take into account the shape of the front. If we wish to compare two fronts we can use the *Hypervolume* metric Guerreiro et al. (2020). The Hypervolume measures the area encompassed by a non-dominated front as shown in Fig. 4.5. As we are considering a problem with 2 solution characteristics, sometimes referred to as a *2-dimensional* problem. When working in 2 dimensions the hypervolume is reasonably simple to comprehend, it measures the area enclosed by the non-dominated front. Figure 4.5 shows a front and the area enclosed, which is defined by the front and the *nadir point*. In Fig. 4.5 the Nadir point would be at (5, 6), which represents the worst values found for characteristics A and B (Fig. 4.6).

If the problem has 3 solution characteristics then the hypervolume will be calculated based on a 3-dimensional space, the hypervolume metric can be used for any number of dimensions which makes it useful when considering non-dominated fronts for multi-objective problems.

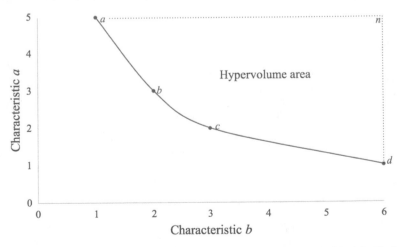

Fig. 4.5 The Hypervolume area bounded by the non-dominated front a, b, c, d and the Nadir point n

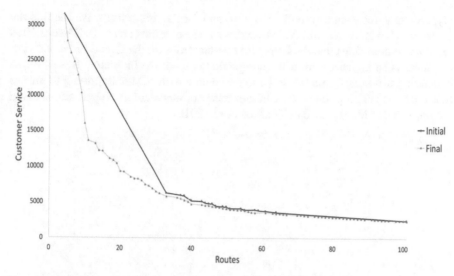

Fig. 4.6 The initial and final non-dominated fronts based on P-n101-k4. For clarity the extreme solutions are Initial Population: (4, 31676.33) and (101, 2494.71) Final Population: (4, 31644.8) and (101, 2494.71)

The larger the Hypervolume, the greater the proportion of the solution space dominated by the front, but what does this mean for ACE Doughnuts? If two non-dominated fronts based on the same problem instance are compared, the one with the greater Hypervolume contains the better solutions. Table 4.5 shows the

Table 4.5 The hypervolume of each individual front and the grand front from the B-n78-k10 instance, the grand front is 2.2% greater than the average

Run	Hypervolume
1	584296.21
2	585481.54
3	588100.99
4	585150.21
5	586476.93
6	586733.97
7	586855.57
8	587436.32
9	588290.44
10	588323.30
Avg.	586714.55
Max.	588323.30
GF Hyper	599124.83

Hypervolume for each run (on B-n78-k10) and for the grand front. We see that the grand front has an increase of 2.2% over the average obtained on the 10 EA runs. This increase demonstrates that the Hypervolume increases on the Grand Front showing that there is an improvement when compared to any of the individual fronts. When comparing different fronts using the hypervolume metric, it is important to ensure that a common nadir point is used. In this instance, we calculate Hypervolume using the implementation released by Fonseca et al. (2010).

```
 1      public BioObjectiveIndividual ( CVRPProblem prob) {
 2          super ();
 3          /*
 4           * Constructor to create a new random genotype
 5           */
 6          boolean all1s = false;
 7          boolean all0s = false;
 8
 9          //Create some with all newVans set to 1 and some with
        all set to 0
10          if (rnd.getRnd().nextFloat() > 0.8){
11              if (rnd.getRnd().nextBoolean()){
12                  all1s = true;
13              }else{
14                  all0s = true;
15              }
16          }
17          problem = prob;
18          genotype = new ArrayList<Gene>();
19          for (Visit v : prob.getSolution()){
20              Gene g = new Gene((VRPVisit)v);
21              if (all1s == true)
22                  g.setNewRoute(true);
23              else if (all0s == true)
24                  g.setNewRoute(false);
25              else
26                  g.setNewRoute(rnd.getRnd().nextBoolean());
27              genotype.add(g);
28          }
29          genotype = randomize(genotype);
30          phenotype = null;
31      }
```

Listing 4.7 The Gene class (within BiObjectiveIndividual.java)

4.7 Conclusions

The approach outlined in this chapter allows ACE Doughnuts to make an informed choice as to which solution to adopt. We recognise that the factors influencing such choices are not always easy to model and allowing a human expert to make final decision may represent the best approach for a real-world problem.

It may be argued that in a scenario such as this, the optimisation is not so much solving the problem as filtering the solutions. If we consider that the starting point of a CVRP problem (or any NP-Hard problem) is a vast number of solutions that we cannot hope to search through using any form of exhaustive search, then the approach

Table 4.6 The set of solutions that comprise the grand front for A-n64-k9

Routes	Customer service	Routes	Customer service	Routes	Customer service
9	14691.63	29	6429.22	52	4360.72
10	13448.21	30	6101.75	53	4152.55
11	13036.44	32	5927.05	56	4096.42
12	11934.61	33	5728.70	57	4038.34
13	11698.51	34	5662.65	58	4026.24
14	11395.04	35	5591.21	59	3944.67
15	10960.03	36	5474.53	60	3902.84
16	9801.78	37	5337.05	61	3900.46
17	9751.78	38	5283.35	62	3859.91
18	9474.77	39	5184.46	63	3816.04
19	8979.62	40	5071.66	64	3816.04
20	8633.49	42	5006.55		
21	8259.49	43	4970.01		
22	7835.90	44	4891.48		
23	7594.50	45	4807.45		
24	7445.42	46	4710.15		
25	7107.62	47	4584.90		
26	6800.84	49	4547.57		
27	6779.95	50	4464.51		
28	6569.62	51	4409.07		

outlined may be considered to be filtering out a smaller representative set of solutions that can be examined by an expert. If we take a typical grand front (see Table 4.6), we note that the user has solutions presented that show (almost) contiguous values from the minimum number of routes (9) to the maximum (1 per customer—64 routes). The planner at ACE Doughnuts can now select a solution that matches their requirements. If we suppose that ACE Doughnuts have a fleet of 20 vehicles available, the planner can see the effect of operating 19 routes (a 4% increase in the customer service cost) should there be an issue with staff or vehicle availability. The planner can also use this technique for answering questions such as "what are the benefits on increasing or decreasing fleet capacity?" or "What do we have to do to decrease our customer service cost?".

The combining of non-dominated fronts produced over individual runs into a grand front opens up the possibility of creating the grand front based on the output of more than one technique. For some problem domains, it may be beneficial to use a range of solvers (or the same solver with differing parameters) in order find solutions at particular points within the solution space.

The data presented in Table 4.6 can be presented in a number of ways which will aid the user in making the final selection. For a bi-dimensional problem, a set of drop-down boxes which allow one objective to be selected and then the corresponding objective to be shown may be used, as can a 2-D plot. By making such interfaces interactive the user could also be presented with the actual solution overlaid onto a map. In the next chapter, where we look many-objective problems we will examine the *parallel coordinates* visualisation technique.

References

Augerat, P. 1995. *Approche polyèdrale du problème de tournées de véhicules. (Polyhedral approach of the vehicle routing problem)*, Ph.D. Thesis. France: Grenoble Institute of Technology. https://tel.archives-ouvertes.fr/tel-00005026.

Fonseca, C. M., Lopez-Ibanez, M., Paquete, L., Guerreiro, A. P. 2010. Computation of the Hypervolume Indicator. http://lopez-ibanez.eu/hypervolume.

Guerreiro, A. P., Fonseca, C. M., Paquete, L. 2020. *The Hypervolume Indicator: Problems and Algorithms*. arXiv:2005.00515.

Vilfredo Pareto | Italian economist and sociologist. 2021. https://www.britannica.com/biography/Vilfredo-Pareto.

Chapter 5
Solving Problems That Have Multiple Solution Characteristics

Abstract In this chapter, we explore strategies for solving problems that have solutions with more than characteristics of interest to the user. In the domain of Delivery Problems, such characteristics could include financial costs, vehicle usage, environmental impact, etc. We previously (Chap. 4) discussed the use of Pareto dominance based to determine non-dominated fronts of solutions. This technique can be expanded to fronts of multiple dimensions, where each dimension represents one solution characteristic. We examine a case study based around the Vehicle Routing Problem with Time Windows (VRPTW) variant of the Vehicle Routing Problem (VRP). VRPTW adds additional time constraints to the basic VRP formulation that require deliveries to be made within specified a time window.

5.1 Last Mile Deliveries

The growth in the on-demand economy (Jaconi 2014) has dramatically increased business to consumer (B2C) deliveries of goods and services to the homes of consumers. Data released by the United Kingdom Office of National Statistics (ONS) (Office of National Statistics 2021) suggests that online purchases have increased from 7.7% of UK retail sales in July 2011 to 26% in July 2021 and reaching 36% in December 2020. The majority of these purchases will be accompanied by a delivery directly to the home of the purchaser. The final step in the supply chain to the home of the consumer is often referred to as the *last mile* (CILT(UK) 2018).

When solving a typical last mile problem, the factors that may have to be taken into consideration include

- **Financial**—capital costs, running costs, energy costs, staff costs;
- **Environmental**—CO_2 emissions, congestion;
- **Customer Service**—delivery time windows;
- **Delivery Technologies**—Vans, Cargo Bikes, Walking couriers, etc.

© Springer Nature Switzerland AG 2022

N. Urquhart, *Nature Inspired Optimisation for Delivery Problems*,
Natural Computing Series, https://doi.org/10.1007/978-3-030-98108-2_5

Such multi-objective problems may be solved in a number of ways; the objectives may be combined by using weights (see Sect. 4.1 to create a single value to be optimised. An alternative approach, as described in the last chapter, is not to attempt to find a single, optimal, solution, but rather to find a set of solutions, each of which is significant, and allow an expert user to make the final choice of solution.

5.2 Vehicle Routing with Time Windows

Many of the home delivery services offered by supermarkets and other providers allow the customer to specify that the delivery is to be made within a given time window.

The generic name for this variant of the VRP is the Vehicle Routing Problem with Time Windows (VRPTW) (Solomon 1987), as with most of the problems examined in this book there exists many variants of the basic problem. The problem that we will examine in this chapter includes a vehicle capacity constraint, so strictly speaking it is the Capacitated Vehicle Routing With Time Windows (CVRPTW) variant. A brief, but by no means exhaustive list of common VRP variants could include

- **CVRP**—Capacitated Vehicle Routing Problem.
  Each vehicle used has a fixed capacity, a vehicle can only visit as many customers as its capacity will allow.
- **VRPTW**—Vehicle Routing Problem with Time Windows (Solomon 1987).
  Each customer must be visited within a specified hard time window.
- **CVRPTW**—Capacitated Vehicle Routing Problem with Time Windows.
  Each customer has a time window constraint and each vehicle has a capacity constraint.
- **SVRP**—Stochastic Vehicle Routing Problem (Cook and Russell 1978).
  Elements of the problem, such as travel times/distances are made variable to better reflect real-world conditions.
- **DVRP**—Dynamic Vehicle Routing Problem (Gendreau and Potvin 1998).
  Customers are added to the problem at run time.
- **VRPSTW**—Vehicle Routing Problem with Soft Time Windows (Kabcome and Mouktonglang 2015).
  Customers are allocated a soft time window (e.g. 10 am to 11 am) and within that a hard time window (e.g. 10:30–10:35). If the delivery takes place within the hard time window no penalty is applied, if the delivery takes place in the soft time window a penalty is applied.
- **VRPPD** Vehicle Routing Problem with Pickup and Delivery (Desaulniers et al. 2002).

Each delivery has pickup and delivery locations. Vehicles must be scheduled to pickup goods and then deliver them. Sub-variants can encompass time windows or heterogeneous vehicle fleets.

Within the CVRPTW a set of customers each require a visit, but unlike the vehicle routing problems discussed in Sect. 2 each customer visit must take place within a time window. Each customer c_i has a time window denoted by a start time s_i and a finish time f_i. This results in the following constraint:

$$s_i <= d_i <= f_i$$

The delivery time of customer i must fall within the time window of customer i. Any solution that has a delivery to i outwith the time window is considered an invalid solution. If the delivery vehicle arrives before the time window, then it may wait for the window to start, but if it arrives later then the delivery cannot be made (many solvers will at this point at the customer to a new route). We must now consider time within our problem formulation and solutions.

5.3 A Case Study: A Home Delivery Service

Fab Foods is a supermarket that decides to trial a home delivery service. Fab Foods' customers can place their orders online and have them delivered to their homes the following day. In common with most services of this type Fab Foods intends to let customers specify a time window when their delivery should be made. The question at this stage for Fab Foods is, how long should that time window be?

As with previous chapters, we make use of the Augerat problem instances (Augerat 1995). As these instances do not have time windows, we generate a set of files with random time windows which are used as follows. We assume that delivery slots can be requested between 08:00 and 22:00. Using the Java program *TWGenerator* (which is available in the accompanying source code repository) we create a set of .csv files which contain a list of randomly generated time windows, with start and end times within the 08:00 to 22:00 period. We generate sets of time windows of 1, 2, 3 ...14 h duration (note that the 14 h time window allows for delivery at any time between 08:00 and 22:00). The *VRPTWProblemFactory* class generates instances of VRPTW problems, by specifying a combination of the Augerat problem instance combined with the time window length. As the Augerat problem instance is loaded, time windows are allocated to customers on the order that appears within the CSV file. In this way, *VRPTWProblemFactory* can generate problem 2036 instances based on the 74 Augerat instances to provide the topology (layout of customers) and demand and the 14 sets of problem windows. Shopping is delivered in plastic crates each of which can contain a selection of grocery items and can be stacked in racks within the delivery vehicles. The demand within the Augerat problems is taken to be the number of crates to be delivered to each customer.

Fab Foods have four objectives which they wish to take into account:

- **Distance**—The total distance of all of the routes within the solution.
- **Time**—The sum of the length of time for each route.
- **Routes**—The total number of routes within the solution.
- **Cost/Crate**—The cost for each Crate delivered (see below).

FabFoods have a costing model for their proposed deliveries which allows the cost/Crate to be calculated as follows:

$$vehicleCost = (noRoutes * VEHFIXED) + (totalDistance * VEHRUNNING)$$

$$staffCost = (totalTime * STAFF\_RATE)$$

$$solutionCost = vehicleCost + staffCost$$

$$costPerCrate = \frac{solutionCost}{totalDemnd}$$

In our example the costs are set as follows:
$VEHFIXED = 164$
The fixed cost of including a vehicle in the solution
$VEHRUNNING = 0.117$
The cost per unit of distance of a vehicle
$STAFF\_RATE = 12$
The cost per hour of employing a member of staff

We assume that one vehicle is required per route and one member of staff is required per vehicle. Staff are paid from when the vehicle leaves the depot until when it returns.

```
1  public class FabFoodCostModel {
2      private double VEH_FIXED = 164; // Pounds
3      private double VEH_RUNNING = 0.117;/ Pounds per km
4      private double STAFF_HOUR = 12;//15;
5
6      //Singleton code
7      private FabFoodCostModel() {}
8      private static  FabFoodCostModel instance=null;
9
10     public static FabFoodCostModel getInstance() {
11         if (instance == null)
12             instance = new FabFoodCostModel();
13         return instance;
14     }
15     //Done Singleton
16     public double getStaffCost(MObjTWIndividual i) {
17         double perMin = STAFF_HOUR/60;
18         return (i.getTime()* perMin);
19     }
20
21     public double getFixedVehCost(MObjTWIndividual i) {
22         return i.getRoutes() * VEH_FIXED;
23     }
24
25     public double getVehRunningCost(MObjTWIndividual i) {
26         return i.getDistance()*VEH_RUNNING;
27     }
28
29     public double getVehicleCost(MObjTWIndividual i) {
30         return this.getFixedVehCost(i)+this.getVehRunningCost(i);
31     }
32
33     public double getSolutionCost(MObjTWIndividual i) {
34         return this.getStaffCost(i) + this.getVehicleCost(i);
35     }
36 }
```

Listing 5.1 The Cost model implemented in Java

Representation and Decoding

The same representation as used in Sect. 2.2 is used; the decoding mechanism that translates from genotype to phenotype is slightly more complex that used for the CVRP. As the decoder calculates the arrival time of each customer, they are added to the solution. If the arrival time is early then the vehicle waits at the customer until the customer is ready (known as *waiting time*). If a vehicle arrives at the customer beyond the time window then a new route is started and the customer is added to the start of that route.

For example, let's assume that we have the following set of customers:

ID	Window		Requirement
	Start	**End**	
1	09:00	10:00	5
2	09:00	10:00	5
3	11:00	12:00	10
4	13:00	14:00	10
5	12:00	13:00	5
6	10:00	11:00	5
7	11:00	12:00	5
8	13:00	14:00	5

The travelling times (minutes) between the customers are

	To								
c	c_1	c_2	c_3	c_4	c_5	c_6	c_7	c_8	d
c_1		12	20	10	12	15	25	15	12
c_2	22		6	25	20	25	8	12	20
c_3	20	5		8	20	25	8	5	12
c_4	12	22	25		20	12	8	20	20
c_5	40	15	20	23		15	45	10	12
c_6	12	5	12	8	6		12	40	8
c_7	25	20	12	10	45	12		18	12
c_8	30	8	35	23	4	12	12		23
d	12	10	15	30	8	20	5	8	

(left label: **From**)

where d is the depot and c_1–c_8 are customers. The time taken to make a delivery will be 5 min and the vehicle capacity will be 20. We will assume that the rounds can leave the depot at 08:30.

Let us assume that we have the following genotype:

5,false	2, false	4, false	8, false	1,false	7,true	3,false	6,false

The decoder creates the first route, and adds customer 5 to the head of it. The initial arrival time is 08:38, based on 8 min travelling time, but as customer 5 is not available until 12:00, the arrival time will be scheduled for 12:00, with the departure from 5 being 12:05.

Route 1
5 @ 12:00

In this case, we use the @ symbol as a short form for "arrives at". The next customer for adding to the solution is 2. The earliest arrival time for customer 2 will be 12:20 (depart from 5 at 12:05 plus 15 min travelling time). The last permissible arrival time for customer 2 is 10:00, so it cannot be added to route 1 after customer 5, therefore a new route is required:

Route 1	Route 2
5 @ 13:00	2 @ 09:00

The decoder continues to process the next two genes, further adding to route 2:

Route 1	Route 2
5 @ 13:00	2 @ 09:00
	4 @ 13:00
	8 @ 13:25

As route 2 is now at capacity, next delivery will have to be added to a new route. The decoder processes the rest of the genotype and produces the following phenotype:

Route 1	Route 2	Route 3	Route 4	Route 5
5 @13:00	2 @ 09:00	1 @ 09:00	7 @ 11:00	6 @ 10:00
	4 @ 13:00		3 @ 11:17	
	8 @ 13:25			

Note that route 4 is started as the genome specifies that customer 7 should be at the start of a new route. For completeness, we can add in the depot departure and arrival times for each route (Note that we use the "*" symbol as a short hand for "departs at").

Route 1	Route 2	Route 3	Route 4	Route 5
D * 12:52	D * 08:50	D * 08:48	D * 10:55	D * 09:40
5 @13:00	2 @ 09:00	1 @ 09:00	7 @ 11:00	6 @ 10:00
D @ 13:17	4 @ 13:00	D @ 09:17	3 @ 11:17	D @ 10:13
	8 @ 13:25		D @ 11:34	
	D @ 13:53			

There is another layer of optimisation; it is possible for a single vehicle to operate Route 3 then Route 5 and then Route 4 and finally Route 1 assuming that there is sufficient time to re-load the vehicle in between runs.

It is worth noting that this simple problem formulation is missing a number of constraints such as drivers' hours which may limit the amount of time a driver can operate a vehicle without a break. Such rules may be based on legal requirements or on local employment policy, they normally take the form of x minutes break must be taken after y minutes driving. It is a relatively simple operation to modify the decoder to take account of any such constraints.

Implementation

Algorithm 13 describes our overall approach to the problem, we solve it first as a single-objective solution and then as a multi-objective problem.

Algorithm 13 Home Delivery Solver

1: **Procedure** HomeDeliverySolver(problem)
2: $solution Pool = []$
3: **for** $tries = 1 to 10$ **do**
4:     $solution Pool.add(single Objective(problem, costCrate))$
5:     $solution Pool.add(single Objective(problem, distance))$
6:     $solution Pool.add(single Objective(problem, costTime))$
7:     $solution Pool.add(single Objective(problem, Routes))$
8: **for** $tries = 1 to 10$ **do**
9:     $solution Pool.add All(multi Objective(problem))$
10: $non Dom = extract Non Dom(solution Pool)$
11: **Return** $non Dom$

As we are using object-oriented techniques when implementing our algorithm, we can implement the scheme outlined (Algorithm 13) by making maximum use of code that we have written previously. Most of the changes required to accommodate time windows are contained within the decoding mechanism of the individual, as it is at this point that the time window constraints affect the construction and evaluation of the solution. We extend the *EAIndividual* class and add the additional features to allow time windows to be included.

We implement two solvers. *VRPTWEa* is a single-objective EA (see Listing 5.2) which extends *Solver* and we reuse the *NonDomEA* class that we developed in Chap. 4 (see Listing 4.6).

The java implementation of Algorithm 13 is given in Listing 5.3.

```
1   public void solve() {
2       //Initialise population and keep track of best individual
        found so far
3       bestSoFar = InitialisePopution();
4       while(evalsBudget >0) {
5           //Create child
6           MObjTWIndividual child = null;
7           if (rnd.getRnd().nextDouble() < XO_RATE){
8               //Create a new Individual using recombination,
        randomly selecting the parents
9               child = new MObjTWIndividual(super.theProblem,
        tournamentSelection(TOUR_SIZE),tournamentSelection(TOUR_SIZE
        ));
10          }
11          else{
12              //Create a child by copying a single parent
13              child = (MObjTWIndividual) tournamentSelection(
        TOUR_SIZE).copy();
14          }
15          child.mutate();
16          child.evaluate();
17          evalsBudget --;
18          //Select an Individual with a poor fitness to be
        replaced
19          MObjTWIndividual poor = tournamentSelectWorst(
        TOUR_SIZE);
20          if (poor.evaluate() > child.evaluate()){
21              //Only replace if the child is an improvement
22
23              if (child.evaluate() < bestSoFar.evaluate()){
24                  bestSoFar = child;
25              }
26              population.remove(poor);
27              population.add(child);
28          }
29      }
30      super.theProblem.setSolution((ArrayList)bestSoFar.
        getPhenotype());
31      ((VRPTWProblem)super.theProblem).best = bestSoFar;
32  }
```

Listing 5.2 The main loop of the single-objective EA

```
1    private static NonDominatedPop run(String path,String
     probName,int winLen) {
2         //Solve the instance named in  <probName>
3         VRPTWProblem myVRP = VRPTWProblemFactory.buildProblem(
     path, probName,winLen);//Load instance from file
4         NonDominatedPop solutionPool= new NonDominatedPop();//
     Create an empty population
5
6         //Solve using the Evolutionary Algorithm 4 times, once
     for each objective
7         //1 . cost_del
8         MObjTWIndividual best = solve(myVRP,MObjTWIndividual.
     Objective.COST_DEL);
9         solutionPool.add(best);
10        //2 . routes
11        best = solve(myVRP,MObjTWIndividual.Objective.ROUTES);
12        solutionPool.add(best);
13        //3 . time
14        best = solve(myVRP,MObjTWIndividual.Objective.TIME);
15        solutionPool.add(best);
16        //4 . dist
17        best = solve(myVRP,MObjTWIndividual.Objective.DISTANCE);
18        solutionPool.add(best);
19
20        //now run pareto front
21        String pSizeS = probName.split("-")[1];
22        pSizeS = pSizeS.replace("n","");
23        int size = Integer.parseInt(pSizeS);
24        double avgBags = myVRP.getTotalDemand()/size;
25        for (int x=0; x < 10; x++){
26            NonDomEA eaSolve = new NonDomEA();
27            myVRP.solve(eaSolve);
28            solutionPool.addAll(eaSolve.getNonDom());
29        }
30        //extract GrandFront
31        return  solutionPool.extractNonDom();
```

Listing 5.3 The Java implementation of the home deliveries system

Single-Objective Results

Initially, we consider the problem as a single-objective problem, focusing on distance. As we are using benchmark problems to represent the Fab Foods data, the optimal number of routes are known for each instance; therefore, we can highlight the increase in routes as a result of imposing time windows.

We solve the problem by treating it as a single-objective problem; to address all 4 objectives we solve each instance 4 times, once for each objective. Tables 5.2, 5.3 and 5.4 summarise the results obtained when formulating the problem as a single-objective problem.

A summary of the costs per Crate is given in Table 5.1. We note that as a general rule the cost of allowing customers to specify a 1 h time window effectively doubles the costs of delivery. We would expect that optimising on the delivery cost/Crate would result in the lowest costs, but optimising on distance (with the 14 h window)

Table 5.1 A summary of the average delivery cost per Crate when optimising on each of the 4 single objectives

	Cost/Crate		Route		Time		Dist	
	1 h	14 h	1 h	14 h	1 h	14 h	1 h	14 h
A	6.99	3.67	7.99	4.09	8.33	3.70	7.91	3.56
B	7.21	3.64	8.18	4.11	8.47	3.62	7.84	3.45
P	4.41	2.40	5.02	2.54	7.32	2.36	5.25	2.47

sometimes results in lower costs. The cost is largely based on distance and distance having a very close relationship to how the deliveries are arranged in the chromosome. From a users' perspective, this is partially counter-intuitive (especially if the user is not familiar with the costing model) as it could be expected that optimise on cost would result in the production of solutions with the lowest costs.

From the perspective of the user, treating this problem as 4 separate single-objective problems has the disadvantage that although the user has 4 solutions from which to choose for any problem instance; these solutions represent the extremes of the non-dominated front and don't showcase any possible trade-off solutions.

5.4 Finding Trade-Offs Using Pareto Dominance

As discussed above, the single-objective optimisations produce extreme solutions that prioritise one objective at the expense of the others. We can use a 4-dimensional version of the Pareto Dominance-based algorithm (discussed in Chap. 5), this produces a non-dominated front of solutions for each problem.

As previously, we execute the multi-objective algorithm 10 times and combine the results into a single grand front. We can go further and add the initial extreme results as well. From an implementation perspective, this means that we are executing 10 runs of the multi-objective algorithm and 10 runs of each of the single-objective algorithms (50 executions in total). Whilst that may seem wasteful, the reader could consider the following strategies for implementation:

1. Execute all of the algorithms in a linear fashion on one machine.
2. Execute the algorithms in parallel on a cluster (e.g. cloud based).
3. Reduce the number of individual runs that contribute to the grand front, but accept that there may be a decrease in the range covered by the grand front.

From Table 5.5 it becomes apparent that having smaller time windows (e.g. increasing the constraint) increases the choice of solutions. Where a 14 h window is made available then the algorithm can converge towards a solution that optimises many visits into longer routes. This allows the construction of solutions which dominate many others, leading to relatively few within the Pareto front.

Table 5.2 Solving the problems in set A as 4 single-objective problems

Problem	Time windows	Cost/Crate				Route				Time				Distance			
		Cost/Crate	Routes	Time	Dist	Cost/Crate	Routes	Time	Dist	Cost/Crate	Routes	Time	Dist	Cost/Crate	Routes	Time	Dist
A-n32-k5	1	6.49	8	4793	3350.63	7.66	10	5425	3555.70	6.20	10	2748	3000.55	7.81	11	5488	2553.80
A-n33-k5		5.54	8	4313	2528.05	6.73	10	4949	3186.71	7.11	14	2777	2729.03	6.51	11	4360	1953.42
A-n33-k6		4.54	8	4028	2878.41	5.01	9	4443	2955.98	5.44	13	2587	2505.26	6.21	14	4132	2032.57
A-n34-k5		5.44	8	4290	2838.61	6.44	10	4656	3336.67	6.85	13	3285	3076.49	7.01	11	5674	2446.01
A-n36-k5		6.55	9	5192	3241.00	8.01	11	6345	4008.15	10.10	20	3523	4083.13	7.77	12	5681	2834.40
A-n37-k5		6.90	10	4339	2558.79	8.54	12	5793	2983.80	9.66	18	3102	3075.78	8.99	14	5528	2183.07
A-n37-k6		5.38	10	5047	3570.68	6.64	12	6672	4134.37	7.50	19	3699	3580.11	6.48	15	4479	2864.63
A-n38-k5		6.16	10	4603	3438.88	6.89	11	5461	3585.95	7.94	17	3206	3345.28	8.85	16	6668	2562.58
A-n39-k5		6.69	10	5655	3481.81	7.65	12	6165	3716.89	9.01	19	3772	3485.95	8.05	14	5967	2843.05
A-n39-k6		6.35	11	5223	4190.46	7.44	13	6560	4031.22	9.41	23	3616	3880.56	7.91	16	6031	2833.55
A-n44-k7		6.97	13	6676	4323.50	8.10	16	7295	4553.19	9.41	24	4406	4647.10	7.24	14	7173	3408.47
A-n45-k6		7.08	14	6890	4473.97	7.98	16	7783	4703.96	8.89	24	4230	4186.83	8.40	19	7265	3501.49
A-n45-k7		6.59	13	7445	4776.61	7.58	16	7805	5290.02	6.60	16	5161	4536.37	7.20	17	6761	3623.77
A-n46-k7		6.03	12	6015	3998.37	8.24	16	8681	5187.47	6.15	15	4018	3803.90	7.30	16	7063	3123.53
A-n48-k7		7.22	15	7503	4793.31	8.59	17	9610	5723.82	7.84	20	5454	4610.72	7.87	17	8375	3958.04
A-n53-k7		7.58	16	8535	6011.64	8.76	20	9160	5999.97	8.12	22	5919	5107.02	7.99	20	7666	4222.28
A-n54-k7		8.28	19	8872	5551.31	9.37	21	10594	6029.73	9.99	29	6089	6089.47	8.79	22	8610	4728.74
A-n55-k9		5.72	17	6992	5245.43	7.60	22	10740	5326.10	8.90	35	5149	5942.79	6.72	22	7849	3915.12

(continued)

Table 5.2 (continued)

Problem	Time windows	Cost/Crate				Route				Time				Distance			
		Cost/Crate	Routes	Time	Dist	Cost/Crate	Routes	Time	Dist	Cost/Crate	Routes	Time	Dist	Cost/Crate	Routes	Time	Dist
A-n60-k9		7.57	21	10185	6756.72	8.30	23	11145	7515.10	6.93	22	7225	5904.74	8.00	24	10245	5560.39
A-n61-k9		6.94	23	8877	5053.64	7.19	23	9578	5744.61	9.86	42	5934	5553.82	7.91	28	9640	4098.65
A-n62-k8		9.45	23	11266	7721.03	9.78	24	11835	7422.60	8.01	22	7477	6551.64	9.73	26	10808	6044.91
A-n63-k10		6.85	21	10992	6371.59	7.65	25	11225	6716.98	9.14	39	6785	6579.54	7.43	26	10438	4871.94
A-n63-k9		8.07	24	10890	7968.90	9.19	25	14033	9553.94	6.93	21	8474	7751.62	8.14	24	11631	7197.55
A-n64-k9		7.80	22	11028	6841.93	8.89	25	12710	7668.60	7.40	24	8032	6281.90	8.33	24	11779	6567.09
A-n65-k9		8.13	25	10887	7319.38	8.31	26	10937	7150.16	10.36	42	7031	6748.06	7.52	24	10272	5133.73
A-n69-k9		8.41	26	10148	6927.78	8.94	27	11396	7223.87	11.03	43	7113	7231.85	8.77	28	10987	5294.20
A-n80-k10		9.89	30	15814	10506.71	10.20	31	15867	11579.74	10.18	36	12182	10645.99	10.64	34	16253	10231.13
A-n32-k5	14	3.80	5	2203	2557.11	4.86	5	3452	4112.92	3.68	5	2099	2304.66	3.64	5	2014	2289.56
A-n33-k5		3.40	5	2074	2419.89	4.04	5	2883	3462.20	3.04	5	1621	1806.91	2.95	5	1510	1647.33
A-n33-k6		3.06	6	2046	2260.03	3.54	6	2753	3243.93	2.80	6	1633	1767.02	3.46	8	1728	1824.29
A-n34-k5		3.49	5	2329	2720.27	3.89	5	2860	3377.51	3.48	6	1881	2037.44	3.07	5	1827	1958.00
A-n36-k5		3.81	5	2572	2986.82	4.39	5	3299	3930.02	3.66	5	1969	3443.86	3.38	5	2040	2278.84
A-n37-k5		3.91	5	2311	2640.36	4.59	5	3083	3698.22	3.50	5	1830	2033.07	3.41	5	1711	1936.90
A-n37-k6		3.34	6	2746	3172.16	3.57	6	3155	3612.23	3.86	8	2279	3691.96	3.27	7	2166	2408.29
A-n38-k5		3.56	5	2624	3122.85	3.63	5	2739	3226.12	2.93	5	1776	2015.03	3.41	6	1976	2218.25
A-n39-k5		3.67	5	2730	3208.56	4.32	5	3614	4363.50	4.14	7	2061	3481.36	3.24	5	2191	2419.85
A-n39-k6		3.66	6	2821	3226.48	4.24	6	3666	4365.11	3.33	6	2296	2636.52	3.32	6	2305	2557.95

(continued)

Table 5.2 (continued)

Problem	Time windows	Cost/Crate				Route				Time				Distance			
		Cost/Crate	Routes	Time	Dist	Cost/Crate	Routes	Time	Dist	Cost/Crate	Routes	Time	Dist	Cost/Crate	Routes	Time	Dist
A-n44-k7		3.56	6	3090	3652.97	4.01	6	3867	4523.77	4.71	10	2713	4285.90	3.15	6	2435	2757.09
A-n45-k6		3.85	7	2953	4663.07	3.71	6	3185	4940.13	3.61	7	2931	3477.98	3.39	7	2622	2903.01
A-n45-k7		3.57	7	3370	3792.35	3.95	7	4034	4678.08	3.65	7	3048	4732.05	3.45	7	3110	3574.03
A-n46-k7		3.73	7	3272	3809.61	4.15	7	3976	4789.25	3.47	7	2817	3239.50	3.90	9	2666	2929.07
A-n48-k7		3.84	7	3731	4355.25	4.39	7	4748	5552.68	3.94	7	3485	5319.34	5.14	13	3298	3635.86
A-n53-k7		3.77	7	4003	4716.47	4.13	7	4686	5634.14	3.92	8	3427	5168.24	3.92	9	3417	3807.90
A-n54-k7		3.89	7	3863	5804.23	4.45	7	5354	6456.48	4.06	8	3759	5556.94	3.83	8	3765	4236.46
A-n55-k9		3.39	9	4031	4803.81	3.71	9	4816	5752.48	3.04	9	3223	3684.09	3.01	9	3170	3546.85
A-n60-k9		3.77	9	4928	5689.45	3.98	9	5445	6287.50	3.77	10	4003	5883.54	3.72	10	4356	4913.70
A-n61-k9		3.36	10	3959	4605.49	3.48	9	4695	5681.52	3.11	10	3345	3763.09	3.11	10	3380	3718.77
A-n62-k8		4.12	8	4572	6752.77	4.58	8	6016	7220.24	4.32	9	4192	7303.21	3.94	8	4700	5436.17
A-n63-k10		3.45	10	4677	5485.58	3.75	10	5459	6492.28	4.22	15	4434	5018.13	3.52	12	3949	4442.76
A-n63-k9		3.94	9	5805	6844.71	4.38	9	6957	8174.96	3.99	10	5075	7058.82	3.90	10	5262	6080.92
A-n64-k9		3.60	9	4334	6061.27	3.93	9	5491	6459.46	3.88	10	4502	6439.70	3.88	10	4948	5656.00
A-n65-k9		3.70	10	4347	6255.42	4.11	9	6245	7491.46	3.67	11	4250	4789.10	3.38	10	3992	4507.05
A-n69-k9		3.82	9	4732	6879.38	4.10	9	5858	7002.71	3.78	10	4203	6112.85	3.57	10	4113	4719.62
A-n80-k10		4.15	10	6348	8561.55	4.50	10	7697	9083.83	4.31	11	5897	9221.61	4.04	10	6439	7506.97

Table 5.3 Solving the problems in set B as 4 single-objective problems

Problem	Time windows	Cost/Crate				Route				Time				Distance			
		Cost/Crate	Routes	Time	Dist	Cost/Crate	Routes	Time	Dist	Cost/Crate	Routes	Time	Dist	Cost/Crate	Routes	Time	Dist
B-n31-k5	1	5.42	7	4098	2273.21	7.03	9	5510	2711.45	4.85	7	2879	2348.84	6.62	9	5023	2106.80
B-n34-k5		6.03	9	4763	2795.02	7.35	11	5687	3559.72	5.93	10	3574	3025.65	7.96	13	5908	2754.62
B-n35-k5		6.72	9	5131	3723.88	8.52	11	7078	4321.54	7.45	12	3972	4227.20	6.86	9	5857	2987.64
B-n38-k6		5.96	10	5237	3100.80	7.01	11	6785	3651.31	6.68	14	3438	3738.15	6.16	11	5223	2608.84
B-n39-k5		6.59	10	4588	2933.29	8.53	12	6704	3804.61	10.94	23	3316	3238.04	8.40	14	5618	2373.42
B-n41-k6		6.64	12	6780	3757.40	7.23	14	6553	4204.60	8.04	20	4274	3599.52	6.53	14	5434	2741.29
B-n43-k6		7.06	13	5832	3242.51	7.85	14	6542	4148.96	7.82	18	3805	3096.10	8.77	18	6494	2729.98
B-n44-k7		6.05	13	6324	4096.39	7.21	16	7425	4364.33	6.71	18	4632	3599.87	6.95	16	7285	3175.42
B-n45-k5		8.00	13	6159	4492.22	9.73	16	7777	4696.55	10.72	24	3968	4095.62	7.80	13	6664	2778.32
B-n45-k6		6.27	13	6130	3032.97	7.08	15	6619	3491.53	9.01	25	3955	3769.31	6.75	15	6287	2392.79
B-n50-k7		7.61	16	7661	4083.28	8.31	17	8711	4541.86	10.90	32	4371	4416.85	7.30	16	7400	2946.17
B-n50-k8		6.76	16	8464	5546.24	6.95	17	8620	5100.33	7.28	22	5772	5002.49	6.65	16	8915	4115.46
B-n51-k7		6.28	14	6393	5320.49	8.86	20	9889	6870.34	7.52	21	5222	5572.01	7.80	20	7911	4047.82

(continued)

Table 5.3 (continued)

Problem	Time windows	Cost/Crate				Route				Time				Distance			
		Cost/Crate	Routes	Time	Dist	Cost/Crate	Routes	Time	Dist	Cost/Crate	Routes	Time	Dist	Cost/Crate	Routes	Time	Dist
B-n52-k7		7.58	15	8140	4337.20	8.96	19	8776	4767.74	9.73	27	4466	4909.88	8.50	20	7346	3424.15
B-n56-k7		8.11	18	7786	4161.76	8.62	19	8327	4491.81	12.28	37	5033	4203.30	9.93	25	8094	3376.98
B-n57-k7		8.05	18	9173	7056.57	9.58	22	10823	7713.97	7.73	20	6967	6102..9	8.72	21	10186	5099.05
B-n57-k9		8.30	21	11676	7560.03	8.12	19	12722	7359.43	7.03	19	8579	6937.06	8.51	22	12016	7036.36
B-n63-k10		7.28	24	9647	7222.14	7.88	24	12105	7767.35	7.08	24	8612	7417.19	8.22	27	11768	6778.96
B-n64-k9		6.95	22	9599	4881.03	7.74	25	10125	5719.90	10.13	43	5917	5651.87	7.27	24	10085	3706.80
B-n66-k9		8.18	24	11441	6996.33	8.34	24	11850	7467.64	7.24	24	8010	5923.75	8.12	25	10849	6147.15
B-n67-k10		7.49	24	10702	6145.27	8.30	27	11556	6735.07	10.62	46	6665	6437.39	7.87	28	10167	4344.36
B-n68-k9		9.27	26	12843	7928.26	9.43	27	12650	7964.19	7.94	25	8834	6626.70	9.53	28	13039	6652.47
B-n78-k10		9.33	33	11957	8062.39	9.54	32	13327	8730.10	11.27	48	8766	7986.66	9.13	31	13718	6178.13
B-n31-k5	14	3.29	5	1688	1703.02	3.87	5	2322	2638.61	3.21	5	1556	1647.50	3.20	5	1557	1599.64
B-n34-k5		3.30	5	2083	2333.05	3.72	5	2623	3031.19	3.41	5	1821	3187.97	3.54	6	1969	2057.96
B-n35-k5		4.10	5	2491	4062.36	4.55	5	3474	4048.37	4.23	5	2223	4987.53	3.48	5	2143	2311.70
B-n38-k6		3.52	6	2471	2748.17	4.21	6	3467	4074.80	3.20	6	2020	2147.71	3.18	6	1997	2104.26

(continued)

Table 5.3 (continued)

Problem	Time windows	Cost/Crate				Route				Time				Distance			
		Cost/Crate	Routes	Time	Dist	Cost/Crate	Routes	Time	Dist	Cost/Crate	Routes	Time	Dist	Cost/Crate	Routes	Time	Dist
B-n39-k5		3.65	5	1902	3462.95	4.37	5	3235	3884.44	3.07	5	1644	1726.65	2.97	5	1504	1587.59
B-n41-k6		3.47	6	2930	3396.99	4.06	6	3897	4589.59	3.03	6	2231	2459.99	3.31	7	2251	2375.44
B-n43-k6		3.49	6	2494	2845.89	3.96	6	3187	3756.58	3.99	8	2321	2580.57	3.16	6	2033	2200.37
B-n44-k7		3.38	7	3033	3509.74	3.87	7	3947	4622.15	3.33	8	2502	2768.60	3.37	8	2605	2772.24
B-n45-k5		3.97	5	3285	3878.42	4.30	5	3730	4493.20	3.61	6	2344	2579.07	3.57	6	2274	2528.73
B-n45-k6		3.18	6	2704	3064.84	3.31	6	2891	3400.35	3.72	9	2248	2380.92	3.04	7	2000	2154.10
B-n50-k7		3.73	7	3351	3897.27	4.21	7	4170	4991.85	4.02	10	2474	2688.32	3.08	7	2221	2441.65
B-n50-k8		3.49	8	3795	4253.97	3.88	8	4573	5328.84	3.27	8	3348	3581.51	3.42	8	3634	4034.91
B-n51-k7		3.98	8	3766	5624.34	4.31	7	5292	6325.51	3.75	8	3329	5000.20	3.03	7	2798	3111.94
B-n52-k7		3.65	7	2739	4350.44	4.59	7	4792	5753.31	3.95	7	2854	5757.82	4.15	10	2679	2875.17
B-n56-k7		3.70	7	2958	4595.17	4.19	7	4181	5090.31	3.20	7	2509	2727.54	3.12	7	2356	2595.19
B-n57-k7		4.02	8	4425	5171.85	4.65	8	5706	6742.64	4.33	9	3773	6715.01	3.73	8	3856	4386.96
B-n57-k9		3.76	9	4664	5209.47	4.15	9	5559	6397.81	3.85	9	4839	5524.55	3.80	9	4763	5311.54
B-n63-k10		3.73	10	5384	6176.82	4.16	10	6468	7675.03	4.37	14	4879	6507.92	3.54	10	4860	5548.96
B-n64-k9		3.43	10	4110	4711.53	3.83	10	5089	6056.88	3.03	10	3108	3425.65	2.95	10	2899	3163.02
B-n66-k9		3.65	9	4963	5771.82	3.91	9	5601	6552.52	3.85	10	4547	5534.82	3.73	10	4718	5338.40
B-n67-k10		3.51	10	4570	5343.06	3.96	10	5766	6842.96	3.16	10	3685	4161.37	3.23	11	3413	3769.41
B-n68-k9		3.86	9	5200	6104.91	4.28	9	6209	7408.64	3.83	9	4282	7438.84	3.88	10	4838	5504.29
B-n78-k10		3.85	10	5809	6893.69	4.18	10	6745	7965.54	3.80	10	5242	7418.44	4.79	17	5169	5688.60

Table 5.4 Solving the problems in set P as 4 single-objective problems

Problem	Time windows	Cost/Crate				Route				Time				Distance			
		Cost/Crate	Routes	Time	Dist	Cost/Crate	Routes	Time	Dist	Cost/Crate	Routes	Time	Dist	Cost/Crate	Routes	Time	Dist
P-n101-k4	1	6.73	37	14927	6461.25	7.12	39	16129	6491.66	12.21	94	7680	7266.80	7.42	41	17426	5201.76
P-n16-k8		6.64	8	1115	850.69	7.71	8	2407	884.79	7.89	10	983	898.57	7.38	8	2094	725.70
P-n19-k2		2.85	3	1541	720.81	3.16	3	1888	935.09	3.50	5	867	775.00	3.53	4	1838	602.19
P-n20-k2		2.89	3	1577	764.74	3.66	4	1908	835.71	5.24	8	995	957.93	4.29	5	2148	692.18
P-n21-k2		3.07	3	1699	707.08	3.95	4	2067	930.61	3.81	5	1107	805.63	3.71	4	1862	668.83
P-n22-k2		2.97	3	1621	838.39	4.76	5	2644	1009.52	4.83	7	1116	987.29	5.03	6	2403	713.74
P-n22-k8		0.09	9	2073	1232.91	0.10	9	2568	1467.08	0.11	12	1407	1215.22	0.11	10	3470	1015.19
P-n23-k8		6.74	10	1704	1088.73	7.47	9	3630	1171.91	8.06	13	1272	1176.86	7.32	10	2704	944.21
P-n40-k5		4.42	10	4308	1980.91	4.91	11	4745	2385.99	7.57	24	2407	2239.30	5.33	13	4894	1588.37
P-n45-k5		5.01	13	5117	2650.32	5.15	13	5537	2769.54	8.66	31	2952	2722.93	6.04	17	5859	1857.38
P-n50-k10		4.07	16	4644	2732.44	5.01	18	7392	2863.25	8.13	41	3291	2980.98	4.51	17	6346	2006.29

(continued)

Table 5.4 (continued)

Problem	Time windows	Cost/Crate				Route				Time				Distance			
		Cost/Crate	Routes	Time	Dist	Cost/Crate	Routes	Time	Dist	Cost/Crate	Routes	Time	Dist	Cost/Crate	Routes	Time	Dist
P-n50-k7		4.03	14	6159	2603.75	4.52	17	5916	2766.94	7.42	37	3214	2941.66	4.65	17	6965	2059.41
P-n50-k8		3.93	14	5699	2571.52	4.91	17	7696	2962.35	8.30	42	3243	3081.60	4.64	18	6131	2000.26
P-n51-k10		5.42	16	6284	2821.58	5.84	18	6226	2932.14	9.80	40	3420	3180.52	6.85	21	8033	2292.73
P-n55-k10		4.34	18	6046	3114.15	5.02	20	7913	3184.99	7.36	40	3610	3284.75	5.71	24	8686	2376.42
P-n55-k15		5.09	22	6673	3060.08	5.55	22	8852	3436.62	7.19	39	3509	3407.32	5.95	25	9093	2426.21
P-n55-k7		4.11	16	6643	2826.11	4.99	19	8502	3313.76	7.68	42	3630	3335.13	4.80	19	8029	2353.35
P-n55-k8		4.14	17	6002	2791.01	4.62	18	7526	3025.69	7.53	41	3602	3412.04	5.08	21	7922	2266.79
P-n60-k10		4.70	21	7367	3480.43	5.20	22	9214	3791.47	7.92	47	4089	3882.39	6.00	29	8654	2737.22
P-n60-k15		5.04	22	8471	3488.48	5.49	23	9965	3951.33	8.36	50	4165	3826.06	5.51	25	9131	2752.73
P-n65-k10		4.79	22	8826	3932.71	5.18	24	9458	4143.54	8.01	51	4505	4238.48	5.69	29	9079	3134.35
P-n70-k10		4.67	23	9300	4307.96	5.22	26	10323	4470.38	8.95	62	5073	4836.99	4.99	26	9454	3411.41
P-n76-k4		5.26	27	11047	4635.17	5.45	28	11434	4734.62	7.49	52	5437	5131.05	5.14	27	10789	3659.66
P-n76-k5		4.86	26	9274	4412.76	5.53	29	10993	4986.96	9.56	69	5525	5321.69	6.19	35	11420	3616.34
P-n101-k4	14	1.76	6	4718	5501.75	1.64	5	4215	6208.54	1.53	5	4216	4896.17	2.20	11	4215	4745.00

(continued)

Table 5.4 (continued)

Problem	Time windows	Cost/Crate				Route				Time				Distance			
		Cost/Crate	Routes	Time	Dist	Cost/Crate	Routes	Time	Dist	Cost/Crate	Routes	Time	Dist	Cost/Crate	Routes	Time	Dist
P-n16-k8		6.26	8	742	687.84	6.47	8	877	899.83	6.27	8	742	703.00	6.26	8	742	687.84
P-n19-k2		1.67	2	581	628.01	1.84	2	728	815.18	1.59	2	522	529.20	2.08	3	487	466.62
P-n20-k2		1.70	2	613	658.53	1.82	2	720	791.28	1.61	2	539	543.19	1.55	2	475	483.35
P-n21-k2		1.78	2	621	671.59	2.01	2	821	902.20	1.70	2	558	585.25	1.65	2	515	516.70
P-n22-k2		1.77	2	664	724.71	2.01	2	871	988.45	1.62	2	539	526.97	2.17	3	567	538.33
P-n22-k8		0.08	8	1197	1234.85	0.08	8	1406	1541.24	0.08	9	990	996.74	0.08	9	1012	974.17
P-n23-k8		5.71	9	981	973.79	6.09	9	1313	1439.04	5.66	9	947	913.35	5.65	9	944	898.09
P-n40-k5		2.24	5	1693	1910.33	2.51	5	2161	2538.12	2.05	5	1377	1484.34	2.02	5	1317	1397.81
P-n45-k5		2.11	5	1915	2191.22	2.41	5	2515	2939.33	1.97	5	1637	1847.98	1.95	5	1603	1757.85
P-n50-k10		2.56	10	2398	2663.61	2.60	10	2527	2823.10	2.70	12	1856	1926.21	2.68	12	1830	1864.39
P-n50-k7		1.98	7	2204	2495.74	2.11	7	2569	2979.91	1.80	7	1732	1890.05	1.78	7	1662	1787.43
P-n50-k8		2.32	9	2205	2482.29	2.53	9	2778	3212.44	2.19	9	1845	1996.66	2.34	10	1790	1905.95
P-n51-k10		3.34	11	2400	2633.90	3.40	10	2976	3479.41	3.17	11	2033	2175.58	3.33	12	1928	1985.07

(continued)

Table 5.4 (continued)

Problem	Time windows	Cost/Crate				Route				Time				Distance			
		Cost/Crate	Routes	Time	Dist	Cost/Crate	Routes	Time	Dist	Cost/Crate	Routes	Time	Dist	Cost/Crate	Routes	Time	Dist
P-n55-k10		2.36	10	2476	2761.28	2.59	10	3136	3647.31	2.51	12	1989	2089.90	2.33	11	1927	2044.76
P-n55-k15		3.44	16	2908	3230.04	3.51	16	3118	3501.78	3.53	18	2279	2301.80	4.17	22	2338	2335.34
P-n55-k7		1.86	7	2370	2690.50	1.94	7	2589	3007.71	1.71	7	1928	2124.26	1.66	7	1795	1904.87
P-n55-k8		1.83	7	2268	2573.54	1.98	7	2727	3162.62	2.02	9	1912	2061.93	1.83	8	1823	1957.23
P-n60-k10		2.26	10	2772	3132.82	2.42	10	3291	3798.82	2.53	13	2276	2442.97	2.38	12	2230	2405.66
P-n60-k15		3.14	16	2865	3139.85	3.16	15	3382	3811.81	3.03	16	2509	2620.56	3.29	18	2454	2500.86
P-n65-k10		2.22	10	3207	3638.87	2.28	10	3397	3968.56	2.17	11	2560	2819.33	2.16	11	2530	2763.49
P-n70-k10		2.22	10	3793	4429.03	2.28	10	3993	4760.81	2.09	11	2874	3149.97	2.95	18	2834	2998.73
P-n76-k4		1.48	5	3552	4168.04	1.64	5	4163	4987.67	1.28	4	2806	4484.22	1.17	4	2823	3224.55
P-n76-k5		1.49	5	3573	4202.33	1.57	5	3915	4644.51	1.75	8	2798	4375.01	1.68	8	2973	3329.42

Table 5.5 The sizes of the grand fronts produced by combining the output of the single-objective EAs and the multi-objective EA, for time windows of sizes 1 h and 14 h

Problem	TW Size		Problem	TW Size		Problem	TW Size	
	1	14		1	14		1	14
A-n32-k5	14	2	B-n31-k5	6	1	P-n101-k4	85	2
A-n33-k5	20	1	B-n34-k5	9	2	P-n16-k8	16	1
A-n33-k6	15	1	B-n35-k5	20	2	P-n19-k2	9	3
A-n34-k5	27	2	B-n38-k6	18	1	P-n20-k2	7	1
A-n36-k5	8	4	B-n39-k5	14	1	P-n21-k2	14	1
A-n37-k5	16	2	B-n41-k6	20	4	P-n22-k2	14	2
A-n37-k6	13	4	B-n43-k6	20	1	P-n22-k8	30	4
A-n38-k5	20	1	B-n44-k7	21	6	P-n23-k8	27	2
A-n39-k5	13	2	B-n45-k5	20	4	P-n40-k5	40	1
A-n39-k6	10	3	B-n45-k6	29	2	P-n45-k5	35	1
A-n44-k7	5	3	B-n50-k7	17	1	P-n50-k10	46	4
A-n45-k6	18	2	B-n50-k8	9	1	P-n50-k7	56	1
A-n45-k7	24	1	B-n51-k7	33	2	P-n50-k8	26	2
A-n46-k7	25	1	B-n52-k7	21	3	P-n51-k10	71	3
A-n48-k7	19	3	B-n56-k7	39	1	P-n55-k10	45	3
A-n53-k7	20	4	B-n57-k7	14	3	P-n55-k15	46	4
A-n54-k7	8	7	B-n57-k9	4	3	P-n55-k7	30	1
A-n55-k9	36	1	B-n63-k10	22	3	P-n55-k8	57	3
A-n60-k9	22	3	B-n64-k9	31	1	P-n60-k10	43	2
A-n61-k9	27	2	B-n66-k9	8	3	P-n60-k15	55	2
A-n62-k8	19	4	B-n67-k10	25	2	P-n65-k10	32	4
A-n63-k10	15	2	B-n68-k9	14	5	P-n70-k10	64	2
A-n63-k9	17	3	B-n78-k10	17	9	P-n76-k4	80	2
A-n64-k9	19	2				P-n76-k5	82	3
A-n65-k9	26	1						
A-n69-k9	35	6						
A-n80-k10	10	5						

Table 5.6 The Hypervolume values for the grand front created for each problem including the 14 h and 1 h variants. Note that in every case the hypervolume increases for the 14 h delivery slot. This suggests that the solutions based on a 14 h delivery slot perform far better than those constrained to 1 h slots

Problem	Time Windows		Problem	Time Windows		Problem	Time Windows	
	1	14		1	14		1	14
A-n32-k5	6914925.37	90211910.50	B-n31-k5	626580.73	16943928.60	P-n16-k8	779011.02	2270002.16
A-n33-k5	21411224.55	165261964.39	B-n34-k5	1373133.26	69305517.30	P-n19-k2	548451.32	5386206.60
A-n33-k6	16850145.76	110228054.96	B-n35-k5	1827841.50	352172006.45	P-n20-k2	899673.17	9758862.52
A-n34-k5	33197010.54	341571125.95	B-n38-k6	10527863.68	164859336.59	P-n21-k2	1208400.01	15169884.04
A-n36-k5	7086971.75	183645549.17	B-n39-k5	78119314.52	742028589.90	P-n22-k2	974258.69	13322995.02
A-n37-k5	216783414.63	728663661.01	B-n41-k6	63198908.57	567392511.28	P-n22-k8	1737964.82	4140745.73
A-n37-k6	5751817.00	106798533.91	B-n43-k6	11035462.28	145484583.19	P-n23-k8	2861400.58	12641747.50
A-n38-k5	68438637.40	585797557.87	B-n44-k7	30823745.02	412545171.83	P-n40-k5	120666143.37	433630271.34
A-n39-k5	16846282.91	22-852408.43	B-n45-k5	227227997.65	1429407150.79	P-n45-k5	75410670.59	398402056.17
A-n39-k6	191240522.20	705407274.66	B-n45-k6	34006887.94	282818742.57	P-n50-k10	180491591.76	606825458.17
A-n44-k7	19091062.31	282538880.35	B-n50-k7	380344198.02	2346931377.73	P-n50-k7	166704781.19	688318656.74
A-n45-k6	24462172.55	446566361.98	B-n50-k8	17684884.74	274803471.24	P-n50-k8	122017987.67	478565685.17
A-n45-k7	28486583.18	304533359.93	B-n51-k7	80913418.63	1237458484.65	P-n51-k10	302827561.37	1074205618.80
A-n46-k7	14666814.38	227240791.42	B-n52-k7	1501186492.40	147355180.57	P-n55-k10	206297440.17	900729939.61
A-n48-k7	58111619.51	4705726617.44	B-n56-k7	138707414.84	1678312054.00	P-n55-k15	122658641.20	445873338.53
A-n53-k7	51315028.24	788772066.81	B-n57-k7	78099991.13	1415197827.97	P-n55-k7	221208916.28	1125247608.80
A-n54-k7	919684.00	2180026760.45	B-n57-k9	179095.81	210456495.34	P-n55-k8	294267537.60	1412450846.68
A-n55-k9	388731374.55	2086250047.70	B-n63-k10	7353036.12	500063766.07	P-n60-k10	70780709.73	592887807.74
A-n60-k9	33729597.35	882780843.75	B-n64-k9	213509804.48	1486084700.03	P-n60-k15	123984468.97	578543415.20
A-n61-k9	185176892.67	1136710617.41	B-n66-k9	978808.45	137921569.68	P-n65-k10	333703920.19	1454370582.40
A-n62-k8	145531571.02	671378901.06	B-n67-k10	125933698.30	1687842695.86	P-n70-k10	458585654.14	2130350790.72
A-n63-k10	51018026.76	991026453.03	B-n68-k9	36208047.72	11701420067.49	P-n76-k4	842479911.01	5479268836.07
A-n63-k9	27551857.25	587713349.73	B-n78-k10	1103816377.71	2665343635.26	P-n76-k5	1298800874.29	5605323828.12
A-n64-k9	16109101.58	1137587812.90				P-n101-k4	3505456163.26	23671250499.55
A-n65-k9	602775906.38	4214303618.30						
A-n69-k9	357567159.85	2850098775.01						
A-n80-k10	7116931.69	2628139734.41						

Given that we create 4 extreme solutions using the single-objective EA, we might expect at least these solutions to be present in the grand front. Table 5.5 demonstrates that a front with a size less than 4 may be found. What we are seeing is the ability of the multi-objective EA to find solutions that in some cases dominate some or all of the original 4 solutions. We can illustrate this point if we examine the 4 initial solutions created for problem A-N32-K5, with a 14 h time window:

	Distance	Time	Cost/Crate	Routes
Distance	2289.56	2014	3.64	5
Cost/Crate	2557.11	2203	3.80	5
Routes	4112.92	3452	4.86	5
Time	2304.66	2099	3.68	5

These are the 4 solutions that are initially added to the grand front; the final grand front (after adding the MOEA solutions and extracting the resultant non-dominated sets) is as follows:

Dist	Time	Cost/Crate	Routes	Notes
2289.56	2014	3.64	5	*
2284.09	2079	3.67	5	

*—this solution is the single-objective distance solution.

5.5　Exploring Many-Dimensional Fronts

When we produce a set of solutions for an expert to choose from, there is an argument that states that we have simply replaced a massive problem, that of finding a solution from scratch, with a smaller problem that of selecting a solution from within a pre-defined set. Table 5.7 illustrates the problem.

An alternative view of the data may be obtained by visualising it as a *Parallel Coordinates* plot (Inselberg 2009). Parallel Coordinates are a means of visualising many-dimensional data sets. A vertical axis is drawn for each dimension, and each point is plotted as a poly-line that intersections the axis. Figure 5.1 shows an example of a 4-dimensional point $(5, 8, 2, 6)$ on a PC plot. We can use parallel coordinate plots to visualise groups of solutions, such as the set shown in Table 5.7.

Table 5.7 The Grand Front produced for problem B-n78-k10 with 1 h time windows

	Distance	Time	Cost/Crate	Routes
1	7134.9	8294	10.01	42
2	7329.75	8271	9.86	41
3	6506.86	8539	7.71	29
4	6893.4	8377	8.075	31
5	6738.10	8520	7.74	29
6	6969.36	8346	9.0	36
7	7769.1	8230	10.25	43
8	8042.44	8206	10.28	43
9	6257.1	8573	7.16	26
10	7634.3	8310	9.38	38
11	6525.52	8559	7.19	26
12	6188.65	8804	6.5	22
13	6046.79	9045	6.89	24
14	6029.59	9126	6.73	23
15	6196.21	9045	6.38	21
16	6061.83	8892	6.5	22
17	6057.71	9118	6.55	22

Figure 5.7 demonstrates how an interactive parallel coordinates plot may be used to help understand the solutions contained within a non-dominated front. Figure 5.2 shows all 17 of the solutions in Table 5.7. At this point, the chart is crowded and does not necessarily tell us much. The usefulness of the parallel coordinates plot in our application stems from the ability to apply *filters*. A filter allows us to highlight sections within one or more axis and highlight only those solutions that pass through the filtered area.

If we place ourselves in the position of a manager at FabFoods, then we need to select a solution in order to plan that day's deliveries. Parallel Coordinates provide a means of informing that decision-making process. Let us assume that the manager has been told to reduce the costs of deliveries, but has also had complaints about the length of time that deliveries are taking. These two factors will influence their choice of solution

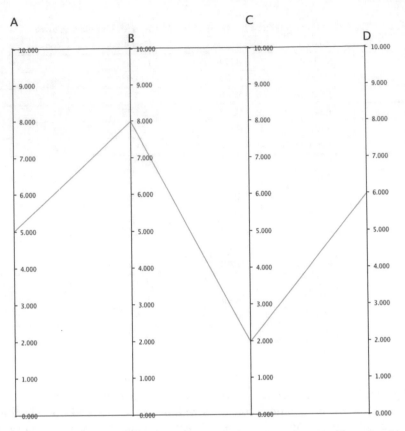

Fig. 5.1 A parallel coordinates plot showing a data point in 4 dimensions. The dimensions$(A - D)$ are represented through the vertical axis. A single point is shown plotted as a poly-line intersecting each of the axis

In order to plot parallel coordinates and to apply filters to them, it is necessary to utilise a package that allows plotting and manipulating of parallel coordinates. In this book, the examples were created using the XDat application (de Rchefort 2020).

If our manager applies a filter to highlight those solutions with lower cost/Crate values, Fig. 5.3 shows the effect of applying a filter at the lower end of the Cost/Crate axis. In this case, we can immediately see that the effect of adopting a solution with a lower cost is to raise the time required. This causes concern as given the complaints that our manager has received about delivery times. A further element to the filter is applied in Fig. 5.4 to the time axis to highlight trade-off solutions that have lower time requirements. Where filters are applied across multiple axis, it is important to remember that the highlighted solutions must pass through all of the elements. Our manager is able to use filtering to find a trade-off solution that goes some way towards satisfying the conflicting demands placed upon the solution.

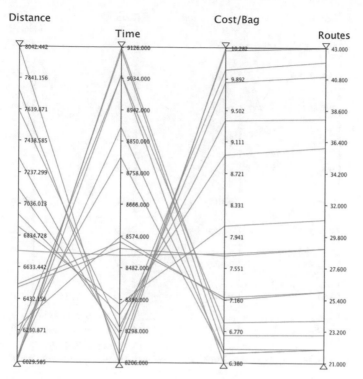

Fig. 5.2 All 17 non-dominated solutions, (based on the data in Table 5.7)

5.6 Conclusions

Within this chapter, we have introduced the concept of having our problem solver produce not one but many possible answers. Evolutionary approaches lend themselves to this approach for two fundamental reasons:

- The are mostly population based, so at any time will have a range of evaluated solutions in memory.
- Being evolutionary they will evaluate many solutions during their execution, potentially covering large areas of the search space.

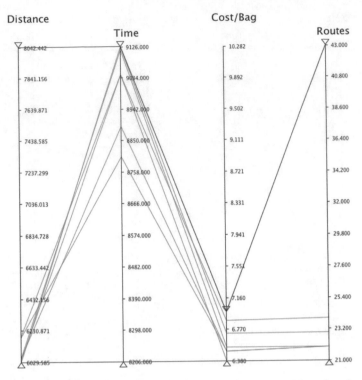

Fig. 5.3 Using filtering to highlight only those solutions with lower cost/Crate, based on Fig. 5.2

When the author first attempted such an approach and created a non-dominated set of solutions to a VRPTW problem and discussed the answer with a colleague from a transportation background, it was pointed out that being presented with a set of 500 solutions to choose from was not really solving the problem from the perspective of the planner. From that conversation stemmed the need for means of supporting the planner in finding the final solution. The interactive parallel coordinates plot has provided a means of visualising a group of solutions supporting the user in making a final selection.

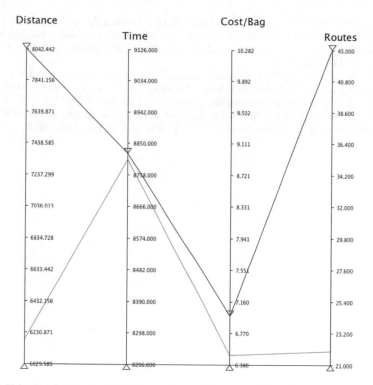

Fig. 5.4 Using the filter to find a the lowest time and lowest cost/Crate, based on Fig. 5.2

References

Augerat, P. 1995. *Approche polyèdrale du problème de tournées de véhicules. (Polyhedral approach of the vehicle routing problem)*, Ph.D. Thesis. France: Grenoble Institute of Technology. https://tel.archives-ouvertes.fr/tel-00005026.

CILT(UK). 2018. The Future of Last Mile Logistics. https://ciltuk.org.uk/News/Latest-News/ArtMID/6887/ArticleID/18808/The-future-of-last-mile-logistics.

Cook, T. M., Russell, R. A. 1978, A simulation and statistical analysis of stochastic vehicle routing with timing constraints. *Decision Sciences* **9**(4): 673–687. https://onlinelibrary.wiley.com/doi/pdf/10.1111/j.1540-5915.1978.tb00753.x.

de Rchefort, E. 2020. XDAT. https://www.xdat.org/.

Desaulniers, G., Desrosiers, J., Erdmann, A., Solomon, M. M., Soumis, F. 2002. 9. VRP with Pickup and Delivery, *The Vehicle Routing Problem*, 225–242. https://epubs.siam.org/doi/abs/10.1137/1.9780898718515.ch9.

Gendreau, M., and J.-Y. Potvin. 1998. Dynamic Vehicle Routing and Dispatching. In *Fleet Management and Logistics*, ed. T.G. Crainic and G. Laporte, 115–126. Boston, MA: Springer, US.

Inselberg, A. 2009. *Parallel Coordinates: Visual Multidimensional Geometry and Its Applications.* Advanced Series in Agricultural Sciences. New York: Springer.

Jaconi, M. 2014. The 'On-Demand Economy' Is Revolutionizing Consumer Behavior–Here's How. https://www.businessinsider.com/the-on-demand-economy-2014-7.

Kabcome, P., Mouktonglang, T. 2015. Vehicle Routing Problem for Multiple Product Types, Compartments, and Trips with Soft Time Windows, *International Journal of Mathematics and Mathematical Sciences* 2015: 126754. Publisher: Hindawi Publishing Corporation. https://doi.org/10.1155/2015/126754.

Office of National Statistics. 2021. Internet Sales as a Percentage of Total Retail Sales (ratio) (%), *Technical Report Source Dataset: Retail Sales Index time Series (DRSI)*, United Kingdom. https://www.ons.gov.uk/businessindustryandtrade/retailindustry/timeseries/j4mc/drsi.

Solomon, M. M. 1987. Algorithms for the vehicle routing and scheduling problems with time window constraints, *Operations Research* 35(2) 35: 254–265 (MAR.- APR., 1987).

Chapter 6
Illuminating Problems

Abstract We have previously examined the technique of combining an Evolutionary Algorithm with Pareto dominance (Chap. 5) to produce a set of solutions to a problem that allows the user to make the final choice of solution. This technique is especially useful in real-world situations where the final choice must take into account external factors (e.g. organisational or political) which influence the desired characteristics of the solution. Within this chapter, we introduce the concept of *illumination algorithms* which produce a range of solutions that illustrate the diversity of solutions that exist within the solution space. The algorithm is only executed once, but the user can then select from a structured set of solutions finding one that, in their opinion, best matches the current requirements. Through a case study based upon supermarket home deliveries, we explore the practicalities and implications of providing user choice within a real-world scenario. We demonstrate the means by which decision to choose a solution can be supported through use of a Parallel Coordinates chart (Inselberg 2009).

6.1 Introduction

It may be argued that the most powerful tool in our quest to solve delivery problems is not computer hardware and software, but the knowledge and experience of a human expert. Techniques such as Evolutionary Algorithms (Holland 1975) allow computers to become effective at the task of searching impossibly large search spaces and evaluating many solutions in a very short time span. If we require a solution that matches a set of well-defined criterion then such approaches will serve us well. However, in real-world scenarios there may exist multiple criterion which will determine the suitability of a solution. Within the context of delivery problems the planner may need to address multiple constraints, stemming from a range of organisational and political objectives and aspirations. The planner may be faced with an organisational commitment to reduce environmental impact by making use of low-carbon delivery modes, but might also be under pressure to reduce operating costs. At that point, the choice of final decision is perhaps best left to an experienced human expert.

© Springer Nature Switzerland AG 2022
N. Urquhart, *Nature Inspired Optimisation for Delivery Problems*,
Natural Computing Series, https://doi.org/10.1007/978-3-030-98108-2_6

In the context of the on-demand economy, such decisions may also have to be taken rapidly as the time scale from a customer placing an order online to expecting delivery grows ever shorter. The time available for running (and re-running) algorithms grows shorter; this becomes acute if a last-minute change of plan is called for due to issues such as staff shortages, weather or vehicle availability.

6.2   Illumination Algorithms—MAP-Elites

As discussed in Chap. 4 the construction of a non-dominated front gives the user a choice of solutions, this has the advantage of incorporating the users' expertise and allowing for a level of decision-making above the planning undertaken by the algorithm. This principle may be extended through the use of an *illumination algorithm* (Mouret and Clune 2015). An Illumination Algorithm seeks to find a set of high-quality solutions that represent the entire solution space, giving the user a large number of solutions to choose from.

The Map Archive of Phenotypic Elites (MAP-Elites) developed by Mouret et al. (2015) seeks to build a *map* of *phenotypic, elites,* in simple terms, the algorithm seeks to build a structured set of high-quality solutions that the user can choose from representing possible solutions from across the entire search space. The key points are that the solutions are held in a structure which allows a user to "browse" through them and that the solutions are *elite*, that is to say they represent high-quality (or in real-world terms, useful) solutions.

Consider a logistics problem that has 2 solution characteristics, financial cost c and environmental impact (CO_2) e. Suppose we evolved the following solutions to our problem:

c	e
500	1023
550	1050
610	990
490	1120

As these solutions have 2 dimensions, we can plot them on a chart.

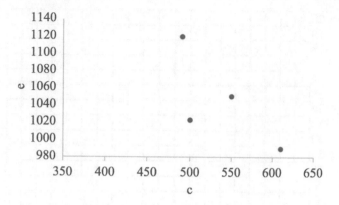

If the ranges for c and e are 350–650 and 950–1140, respectively, and that both values are integers then we could calculate the number of possible solutions as $(650-350) * (1140 - 950) = 57,000$. We could attempt to create an archive of 57,000 solutions and then allow the user to browse through it, but this has a number of practical considerations... ·

- The resources required to maintain a data structure of 57,000 solutions may impose a severe run-time overhead on our softwares.
- Many of these 57,000 solutions may not be valid solutions.
- Many of the solutions may be very similar to others.
- Many of the solutions may be of a poor quality.

And this of course is a very simple example, the space required increases as we add further dimensions onto the problem. MAP-Elites represents the solution space with a smaller archive set of high-quality (or *elite*) solutions. The MAP-Elites archive contains a series of "buckets", every solution in the solution space maps to a bucket. To achieve this, Map Elites scales each solution characteristic into a smaller range. We might decide to scale our two objectives into the range 1–20. It is this set of 400 solutions based on scaled objectives that will form our archive of elite solutions. For example, our empty archive has buckets for 400 solutions:

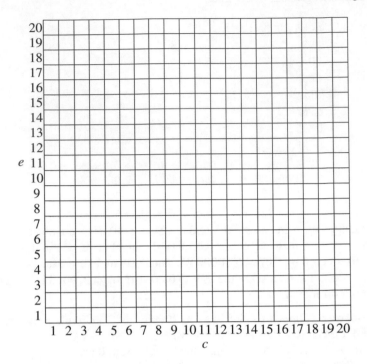

For any characteristic, the raw value r can be translated into the scaled value s as follows:

$$\delta = (max + 1) - min$$

$$cap = \frac{\delta}{b}$$

$$s = int\left(\frac{r - min}{cap} + 1\right)$$

where
max is the maximum value of the characteristic
min is the minimum value of the characteristic
δ is the range of the characteristic
cap is the size of each bin (the number of raw solutions encompassed by each bin)
b is the number of bins.

Using this formula, the four solutions shown above would map to bins as follows:

Raw		Bucket	
c	e	c	e
500	1023	10	8
550	1050	14	11
610	990	18	5
490	1120	10	18

which would occupy the map as follows:

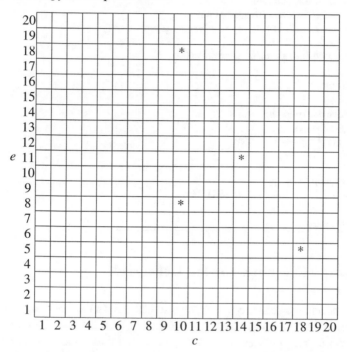

Our example archive of 400 buckets is designed to cover a search space of 57,000 possible solutions. This gives rise to a problem; there are 142 actual solutions that map to each cell. For example, if we generated the solution $c = 512, e = 1020$ that also maps to the cell at 10, 8. There becomes a need to determine which solution may occupy a cell, should two solutions both attempt to occupy it. Within MAP-Elites, we use the concept of a *fitness* value f to determine which solution should occupy each cell.

Our example problem has the solution characteristics (c, e), in addition to this we now need a fitness value f. The fitness value is used to judge the quality of solutions which are competing for the same cell—in practice the fitness function performs the same role that it does in an EA. In this case, we could use distance travelled as our fitness. The overall distance travelled will have a relationship with c and e less distance should equate to less environmental impact and less cost.

Algorithm 14 MAP-Elites

1: **Procedure** $MAP - Elites(init, totEvals, xOverPressure)$
2: $buckets = 20$
3: $dimensions = 2$
4: $evals = 0$
5: $map.initialise(buckets, dimensions)$
6: **while** $evals < init$ **do**
7:     $n = newIndividual()$
8:     $evals + +$
9:     $add(map, n)$
10: **while** $evals < totEvals$ **do**
11:     **if** $random() < xoverPressure$ **then**
12:         $c = newIndividual(map.random(), map.random())$
13:     **else**
14:         $c = newIndividual(map.random())$
15:     $c.mutate()$
16:     $add(map, c$
17: **return** map
18: **Procedure** $add(map, n)$
19: $key = getKey(n)$
20: **if** $map.get(key) == null$ **then**
21:     $map.put(key, n)$
22: **else**
23:     $old = map.get(key)$
24:     **if** $new.fitness() < old.fitness()$ **then**
25:         $map.put(key, n)$
26: **EndProcedure**

6.3   Using MAP-Elites to Plan Deliveries

We will return to the scenario investigated in Chap. 5. In this case, we will examine the problem from the perspective of the planner who must find a solution that best matches the current business objectives and constraints, we will also add cargo bikes into our scenario, the supermarket now having a policy of utilising cargo bikes within their delivery plans. A new plan is required for each working day, the number of deliveries and the quantities will change depending on customer requirements.

A model is required that identifies the costs associated with making deliveries, the values of which are set by the business. The model is applied whilst decoding the genotype ensuring that constraints are taken into account and that the solution is realistically costed. The variables and values used here may be seen in Table 6.2. There is a fixed cost per day associated with each mode, so for each van or bike that cost applies regardless of how much or how little the vehicle is used. The cost per km is applied against the distance travelled (regardless of time taken), the staff cost is applied by time, even if the vehicle is stationary. Vehicles may make several runs per day, loading time refers to the time that it takes to load a vehicle between runs. There are many other variables which may be included in a model such as this, depending on the business requirements.

Table 6.1 The solution characteristics that may be considered by the planner

	Solution characteristic
1	The total daily fixed cost
2	The staff total staff cost
3	The total vehicle running cost
4	The cost per delivery (per crate)
5	The emissions produced
6	The % of deliveries made by bike
7	The % of the distance made by bike
8	The number of bicycles required
9	The number of vans required

Table 6.2 The model for supermarket deliveries

Name	Mode	
	Van	Bike
Capacity	100	20
Fixed Cost (per day)	164	16
Running Cost (per km)	0.117	0.01
Staff (cost per hour)	12	12
Loading time	30	15
Speed	10	1
Emissions	158	0

Our solution characteristics are shown in Table 6.1. Characteristics 1–3 might be paid from differing budgets, so depending on business concerns there may be pressure to reduce some of these costs more than the others. The cost per delivery is noted by the business and is used to compare the efficiency of different branches, our manager may at times feel pressured to minimise this cost. The environmental champion places pressure to reduce emissions (5) and is also keen to see maximum use of cargo bikes (6 and 7). Finally, the number of bicycles and vans at the disposal of the manager will vary from day to day on the basis of staff availability and vehicle maintenance requirements.

A traditional optimisation approach to the above problem might have formulated it as bi-objective based on financial cost versus emissions. But using MAP-Elites, we can optimise taking into account all of the above and then allowing the manager to select a solution based on their requirements.

We use the Augerat instances (Augerat 2014a, b, c) to represent customers and demand. Time windows are generated as described previously (see Chap. 5).

Representation and Decoding

As our problems now encompass two modes of delivery (van and cargo bike), we shall have to modify our representation to take account of the delivery mode. The genotype and decoder presented in Sect. 5.3 is modified with the addition of a delivery mode associated with each gene.

Let us assume that we have a set of customers:

ID	Window		Requirement
	Start	End	
1	09:00	10:00	5
2	09:00	10:00	5
3	11:00	12:00	10
4	13:00	14:00	10
5	12:00	13:00	5
6	10:00	11:00	5
7	11:00	12:00	5
8	14:00	15:00	5

As we have two modes, we must have two sets of travelling times, these are presented in the form *van/bicycle*:

					To					
	c	1	2	3	4	5	6	7	8	d
From	1		12/18	20/31	10/18	12/19	15/23	25/45	15/22	12/22
	2	22/34		6/11	25/41	20/36	25/35	8/16	12/20	20/32
	3	20/30	5/8		8/18	20/35	25/45	8/15	5/9	1/2
	4	12/19	22/41	25/42		20/32	12/18	8/15	20/31	20/35
	5	40/58	15/21	20/39	23/40		15/24	45/71	10/19	12/16
	6	12/18	5/8	12/18	8/17	6/12		12/22	40/51	8/13
	7	25/36	20/30	12/19	10/21	45/60	12/20		18/29	12/20
	8	30/55	8/19	35/46	23/39	4/7	12/19	12/21		2/3
	d	12/18	10/15	15/21	30/42	8/12	20/29	5/8	8/11	

Let us assume that the capacities of each mode are as follows:

- Van = 30
- Cargo bike = 15

Let us assume that the reload time of each mode are as follows:

- Van = 20
- Cargo bike = 10

An example chromosome might look like this:

5,0,V	2,0,B	4,0,B	8,0,B	1,0,B	7,1,B	3,0,B	6,0,B

where each gene contains 3 items,

- **Customer id** 1–8
- **New Route** 0|1
- **Mode** V|B

The decoder creates the first route, and adds customer 5 to the head of it, the mode of a route is determined by the mode of the first customer, in this case the mode of delivery is V (van). The initial arrival time is 08:38, based on 8 min travelling time, but as customer 8 is not available until 12:00, the arrival time will be schedule for 12:00, with the departure from 8 being 12:05.

Routes
1 V
5   12:00

The next customer for adding to the solution is 2. The earliest arrival time for customer 2 will be 12:20 (depart from 5 at 12:05 plus 15 min travelling time). The last permissible arrival time for customer 2 is 10:00, so it cannot be added to route 1 after customer 5, therefore a new route is required, as customer 2 is associated with a B (cargo bicycle) then the new route will be delivered by cargo bicycle:

Routes	
1 V	2 B
5 13:00	2 09:00

The decoder continues to process the next gene adding it to route 2:

Routes	
1 V	2 B
5 13:00	2 09:00
	4 13:00

As route 2 is now at capacity, so the cargo bike must return to the depot, where it arrives at 13:40 allowing for re-loading the bicycle will be available to depart at 13:50.

Routes	
1 V	2 B
5 13:00	2 09:00
	4 13:00
	D

The next delivery for consideration is 8, this can be allocated to route 2, as the bicycle will arrive at 14:01.

The next customer for consideration is customer 1, as they have a 09:00 to 10:00 window they cannot be served on route 2, so another route is started using a van. This continues until the genome has been fully decoded.

Routes			
1V		2B	
5	13:00	2	09:00
		4	13:00
		D	
		8	14:01

Implementation

The multi-modal home deliveries example has been implemented in Java; the source code may be found within the repository that accompanies this book. An outline class diagram is shown in Fig. 6.1.

In order to implement MAP-Elites, we need an efficient implementation of the multi-dimensional archive. Our implementation is contained within the class Archive. The efficiency of this data structure is crucial to the viability of the algorithm, an inefficient structure will result in an algorithm that is too slow to be of practical use.

The capacity of the archive is

$$c = b^d$$

Where
c is the capacity needed
b is the size of the scaled range per dimension
d is the number of solution characteristics (dimensions).

In the FabFoods problem, which has a scaled range of 10 over each of the 9 dimensions, the archive structure needs to have a capacity of 10^9. Attempting to create an array of that length in Java will result in the Java Virtual Machine running out of heap space (although it may be possible to overcome this on some systems by using -*Xmx* option to increase heap space). The storage mechanism adopted here for use within archive is a map which allows items to be added and the tree expanded as required.

The Multi-dimensional Archive

Solutions within the archive are identified by a *key* which comprises an array of integers, one value for each dimension, each item specifying the scaled value of the

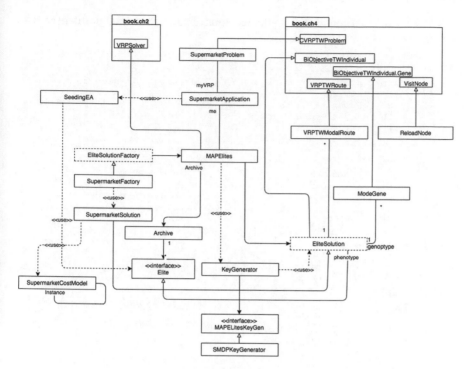

Fig. 6.1 The supermarket delivery MAP-Elites example

solution in that dimension. Each possible key maps to one cell within the archive, so storage and retrieval may be carried out using keys. Objects to be stored in the archive must implement the *Elite* interface, this specifies the requirements for the generation of keys and a fitness value in order to allow the object to be handled by the archive.

The MAP-Elites Algorithm

The MAP-Elites algorithm is implemented as shown in Listing 6.1; we extend the *VRPSolver* class used previously. The implementation of *MapElites* and *Archive* is intended to be reusable for other optimisation problems. The solve() method of our algorithm is shown in Listing 6.1.

In order to use the *MapElites* class a number of other classes must be instantiated:

- *MapElitesKeyGen*—The Key generator produces stores the maximum and minimum values for each solution attribute and generates the appropriate key for a candidate solution.
- *EliteSolutionFactory*—Specifies how new solutions are created, by crossover, cloning or randomisation.
- *EliteSolution*—Specifies the methods that a solution must implement for use with *MapElites*.

Those classes that contain the implementation of the supermarket problem contain *Supermarket* in their names (see Fig. 6.1).

```
1    public void solve() {
2        archive = new Archive(KeyGenerator.getInstance().
     getDimensions(), KeyGenerator.getInstance().getBuckets());
3        initialise();
4        EliteSolution  ch;
5        String descForLogger="";
6        for (int c=0; c < MAX_EVALS; c++) {
7            if ((c%10000)==0) {
8                System.out.print(" "+c + " : "+ archive.size());
9            }
10           Logger.Action action;
11           if (rnd.getRnd().nextBoolean()) {
12               //Select two random parents from the archive
13               EliteSolution p1 = (EliteSolution)archive.
     getRandom();
14               EliteSolution p2 = (EliteSolution)archive.
     getRandom();
15
16               //Create a new solution from the parents
17               ch = (EliteSolution) solutionFactory.getChild(
     theProblem, p1,p2);
18               descForLogger = p1.keyToString() +"," + p2.
     keyToString();
19               action = Logger.Action.RECOMBINATION;
20
21               if (rnd.getRnd().nextBoolean()) {
22                   ch.mutate();            //Mutate the child
     solution
23                   descForLogger = descForLogger + ",MUTATE";
24               }
25           }else {//Create a new solution by copying an existing
     member of the archive
26               ch =  solutionFactory.copy((EliteSolution)archive.
     getRandom());
27               ch.evaluate();
28               descForLogger = ch.keyToString();
29               ch.mutate();
30               action = Logger.Action.CLONE;
31           }
32           ch.evaluate();
33           if (archive.put(ch)) {//Put the solution into the
     Archive. Returns True if this
34               //solution is allowed to join the archive
35               Logger.getLogger().add(action, descForLogger, ch.
     getFitness(),ch.getSummary(), ch.getKey());
36           }
37       }
38       System.out.println();
39   }
```

Listing 6.1   The MapElites algorithm

Setting the Ranges of the Solution Characteristics

As described in Sect. 6.2, each solution characteristic requires a range to be set in order to scale them for the purposes of creating keys. In some cases the setting of the range is straightforward, e.g. if a characteristic is a % (such as items 6 and 7 in Table 6.1) then range is 0 to 100. If we set a range that is outwith the realms of a feasible solution then part of our archive will be unused as it will map to an area of the search space where no feasible solutions exist, and the feasible search space will be mapped by fewer cells. As we shall see it may be difficult for a complex optimisation problem to specify a set of ranges that will only cover feasible areas of the solution space.

One option is to use domain knowledge to specify some of the parameters. For example, it should be possible to create a solution based entirely on one mode of delivery so we know that the range characteristics 8 and 9 (Table 6.1) can begin at 0. But what of the maximum values? One option might be to set them to the number of deliveries allowing for solutions that used 1 vehicle per delivery. Our domain knowledge suggests that many of these solutions with higher number of vehicles are likely to be impractical as there will be constraints placed upon the manager in terms of the vehicles available for use. For these two characteristics, we can reasonably set the range from within the problem instance. The Augerat instances Augerat (1995) specify the minimum number of routes (k) required to solve each instance. In this case, we can use k to determine a reasonable lower bound for each instance. We cannot simply specify k as the upper bound as that would be too constraining. We must allow for the time window constraints and the reduced capacity of the cargo bikes. We must also remember that in order to optimise other solution characteristics it may be necessary to relax the number of vehicles.

To take an example Fig. 6.2 shows a solution space belonging to a problem with 2 solution characteristics. When setting the ranges for x and y we may wish to set the upper range to a value that is at the limit of what we could consider acceptable. Figure 6.2 demonstrates that a reduction in the upper limit of y may exclude solutions which have a lower value of x.

It is not desirable to have to set the ranges manually, many users may lack the knowledge to do so effectively and, as discussed, the use of inappropriate values could prevent desirable solutions from being found (Table 6.3).

We wish to avoid a situation where the bounds have to be specified before a problem instance is solved; we present two possible strategies for this.

- Setting the lower boundaries to 0 and setting the upper boundaries based on the value produced by a randomly generated solution.
- Using a single-objective optimiser on each characteristic to generate a set of realistic lower bound values and the associated upper bounds.

Using the first approach, we set the lower bounds to 0 in each case, we know that for items 1–4 it will not be possible to have a solution of 0 cost, but also we do not know from the problem instance what the lowest cost will be. We set the upper

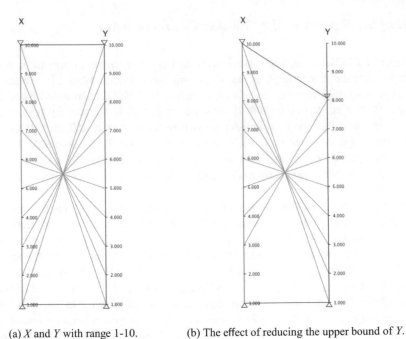

(a) X and Y with range 1-10.   (b) The effect of reducing the upper bound of Y.

Fig. 6.2 We have two solution characteristics X and Y which are opposed. On the left, we see both characteristics with their full range 1–10. Solutions are represented by the green lines linking the two axes. On the right, we see the effect of limiting the range of Y to an upper bound of 8. Observe that limiting the upper bound of Y removes those solutions that perform best with regards to characteristic X

bounds, by generating 10 random solutions and noting the worst result observed for each of the characteristics.

Our second approach makes use of an evolutionary algorithm to optimise based on one specific characteristic and then updates the ranges of each characteristic using the values found across the final population (see Algorithm 15). We will refer to this technique as *seeding*. An advantage of seeding is that the solutions from the final population of each execution of the seeding-ea may be used added to the MAP to initialise it with a range of high-quality solutions. Table 6.4 shows the typical values resulting from the seeded and unseeded procedures. It is worth noting that the seeding process finds higher values for the upper ranges suggesting that the setting of realistic lower values forces the upper bounds to higher values.

Problem Representation and Operators

The representation described in Sect. 6.3 is implemented using the classes *ModeGene* (extends *BiObjectiveTWIndividual.Gene*).The decoder has been implemented in

Table 6.3 The upper and lower bounds for the supermarket delivery problem. Those values marked '?' cannot be determined from an unsolved problem instance

	Solution characteristic	Lower bound	Upper bound
1	The total daily fixed cost	?	?
2	The staff total staff cost	?	?
3	The total vehicle running cost	?	?
4	The cost per delivery (per unit)	?	?
5	The emissions produced	0	?
6	The % of deliveries made by bike	0	1
7	The % of the distance made by bike	0	1
8	The number of bicycles required	0	?
9	The number of vans required	0	?

Algorithm 15 Find Boundaries

1: **Procedure** findBoundaries(problem, objectives)
2: **for each** Objective o : objectives **do**
3:   $sols$:= optimise($o, prob$)
4:   updateRanges($sols$)
5: **EndProcedure**
6: **Procedure** findBoundaries(solutions)
7: **for each** Solution solution : solutions **do**
8:   **for each** Objective objective : objectives **do**
9:     **if** $solution[objective].lower < objective.lower$ **then**
10:       $objective.lower == solution[objective].lower$
11:     **if** $solution[objective].upper > objective.upper$ **then**
12:       $objective.upper == solution[objective].upper$
13: **EndProcedure**

---

the method *SupermarketSolution.decode()*, the principle modification is to allow a vehicle to return to the supermarket and reload. This is achieved by inserting a *ReloadNode* into the Phenotype at the point where the return is required. The presence of the *ReloadNode* ensures that the time and distance calculations allow for a trip to the supermarket, time to reload and then the time required to make the next delivery (see Fig. 6.1).

Table 6.4 An example of the parameters produced using the seeding method (left) and using the non-seeded parameters (right). These parameters are the best found over of 10 runs of seeding EA

Solution characteristic	Seeded				Unseeded			
	Lower	Upper	Delta	Bin size	Lower	Upper	Delta	Bin size
CostDel	1.18	11.9	10.72	1.072	0	7.65	7.65	0.765
CycleDels[a]	0	1	1	0.1	0	1	1	0.1
CycleDist[a]	0	1	1	0.1	0	1	1	0.1
Emissions	0	1230298.8	1230298.8	123029.9	0	659574.8	659574.8	65957.475
Fixed Veh Cost	576	16236	15660	1566	0	10052	10052	1005.2
Veh Rung Cost	73.3	908.7	835.5	83.548	0	536.18	536.18	53.618
Staff Cost	227.4	3613.6	3386.2	338.62	0	3059	3059	305.9
Cycles[b]	0	40	40	4	0	40	40	4
Vans[b]	0	8	8	0.8	0	8	8	0.8

[a]The values for CycleDels and CycleDist are set to 1 and 0 for every problem instance as these values represent percentages
[b]The values for cycles and vans are set based on the value k for that problem instance

The Costing Model

In order to make our software more maintainable, we create a costing model class (*SupermarketCostModel*) which contains the values set in Table 6.2 and any business logic associated with their use. In a production implementation of this system, this class might be implemented to read in the figures from a source such as a spreadsheet to allow for easier updating.

6.4 Results

The Supermarket delivery problem was applied to each of the problems in the Augerat set (Augerat 1995), with 1 h time windows added, each problem was solved 10 times. Each problem was solved using a "seeded" version that set the boundaries of the solution characteristics using the method described in Algorithm 15.

Tables 6.5, 6.6 and 6.7 summarise the results found using MAP-Elites. Whilst the size of the archives makes impressive with an average of 5213 and 2792 solutions produced using the unseeded and seeded variants, respectively. But we should note that the archive has a capacity of 10^9 solutions. In both cases (seeded and unseeded) less than 1% of the buckets in the archive have been filled.

If we consider the average fitness, the seeded solutions have a lower average fitness in every case. This leads to the conclusion, in this case, that the seeding procedure results in a smaller number of solutions for the user to choose from, but they are of a higher quality.

The Managers' Perspective

Let us examine an example from the perspective of the supermarket manager. Orders are placed each day for delivery the following day, there is a relatively short period between the last of the orders being placed and the delivery plan needing to be finalised. If we assume that orders must be placed by midnight then the delivery plan must be finalised in time to allow the first routes to commence at 07:30 the next morning.

A "traditional" approach would allow the algorithm to run for several hours and the resulting plan adopted the following day. Constraints would need to be identified prior to executing the algorithm and it may not be practical to re-run the algorithm if a different solution is required. The MAP-Elites-based approach executes the algorithm once and then allows the manager to make the final choice of solution. This has the potential to allow for last-minute changes which can be accommodated by selecting a different solution from those provided rather than having to execute the algorithm one again.

Table 6.5 A summary of performance on the 'A' problems. Results are given in the format $<$ *unseeded* $>$ / $<$ *seeded* $>$. These results are based on 10 runs, the Size is the number of unique solutions created over 10 runs. The solution characteristics are the lowest value observed during 10 runs

Problem	Size	Avg. fitness	Fixed cost	Cost per Delivery	Staff cost	Running cost	Emissions	% Cycle deliveries	% Cycle dist	Cycles used	Vans used
A-n32-k5-1	5047/2882	5285.5/5179.1	288/304	2.29/2.28	181/142.6	46.59/43.5	0/0	0/0	0/0	0/0	0/0
A-n33-k5-1	6246/3411	4189.5/3941.5	256/352	2.2/2.39	78.8/110.2	34.48/31.96	0/0	0/0	0/0	0 / 0	0/0
A-n33-k6-1	6885/4039	4129.1/4033.6	288/384	1.69/1.66	127.2/117.6	35.08/34.51	0/0	0/0	0/0	0/0	0/0
A-n34-k5-1	4503/2711	4806.3/4666.6	304/384	2.34/2.31	125.6/148.6	40.98/38.47	0/0	0/0	0/0	0/0	0/0
A-n36-k5-1	5029/2619	5613.7/5493.2	256/384	2.35/2.54	198.4/157.8	47.94/45.24	0/0	0/0	0/0	0/0	0/0
A-n37-k5-1	6561/3448	4477.3/4238.3	256/400	2.33/2.62	147.4/129.2	36.2/33.67	0/0	0/0	0/0	0/0	0/0
A-n37-k6-1	5451/3117	5684.4/5609.6	304/400	2.19/2.14	104.8/155.6	48.44/47.53	0/0	0/0	0/0	0/0	0/0
A-n38-k5-1	5947/3021	4867.5/4710.1	336/208	2.34/2.47	91.4/203.8	40.99/39.4	0/0	0/0	0/0	0/0	0/0
A-n39-k5-1	5338/2831	5342.9/5215.8	272/384	2.48/2.73	102.8/136.4	47.28/44.26	0/0	0/0	0/0	0/0	0/0
A-n39-k6-1	6598/4165	5448.1/5274.7	256/432	2.36/2.41	250.8/192.2	46.34/41.98	0/0	0/0	0/0	0/0	0/0
A-n44-k7-1	4896/2487	6343.9/6162.9	320/496	2.42/2.47	117.4/177.0	55.73/54.79	0/0	0/0	0/0	0/0	0/0
A-n45-k6-1	4947/2813	6462.9/6244.3	400/464	2.4/2.48	253.8/233	56.46/53.1	0/0	0/0	0/0	0/0	0/0
A-n45-k7-1	3804/2131	7288.8/7231.3	416/480	2.57/2.64	230.4/230.8	71.61/65.73	0/0	0/0	0/0	0/0	0/0
A-n46-k7-1	5031/2579	6275.3/5997.8	320/432	2.34/2.12	146.6/211	52.57/50.16	0/0	0/0	0/0	0/0	0/0
A-n48-k7-1	4502/2527	7747.4/7600.9	384/496	2.6/2.54	269.8/375	70.02/66.29	0/0	0/0	0/0	0/0	0/0
A-n53-k7-1	4518/2601	7490.5/7326.6	448/528	2.66/2.6	143.2/327.8	66.21/61	0/0	0/0	0/0	0/0	0/0
A-n54-k7-1	4355/2358	8567/8350.8	496/528	2.78/2.65	308.8/264.8	76.88/73.9	0/0	0/0	0/0	0/0	0/0
A-n55-k9-1	5337/2639	7260.2/6991	464/592	2.11/2.25	343.8/185.6	62.99/61.83	0/0	0/0	0/0	0/0	0/0
A-n60-k9-1	4520/2261	9509.2/9181.6	400/608	2.75/2.51	433.2/270.8	87.03/81.48	0/0	0/0	0/0	0/0	0/0
A-n61-k9-1	4398/2622	7470.8/7175	512/628	2.05/2.28	147.8/279.2	63.73/63.56	0/0	0/0	0/0	0/0	0/0
A-n62-k8-1	3753/2207	10176.9/10101.3	576/688	3/3.09	364.6/336.4	101.99/87.82	0/0	0/0	0/0	0/0	0/0
A-n63-k10-1	4568/2302	9125.6/8752.3	544/640	2.34/2.41	255.2/337.6	82.36/76.91	0/0	0/0	0/0	0/0	0/0
A-n63-k9-1	3559/1990	12112.3/11955.2	576/640	2.8/2.92	615.4/438.4	122.03/108.48	0/0	0/0	0/0	0/0	0/0
A-n64-k9-1	3695/2331	10685.7/10466.2	560/592	3.18/2.93	201.4/242.2	100.44/91.78	0/0	0/0	0/0	0/0	0/0
A-n65-k9-1	3698/2079	8903/8719.5	512/624	2.38/2.67	173.2/183.2	80.77/79.09	0/0	0/0	0/0	0/0	0/0
A-n69-k9-1	3955/2380	8498.3/8357.4	592/672	2.58/2.65	229.6/340.4	76.89/73.63	0/0	0/0	0/0	0/0	0/0
A-n80-k10-1	2659/1802	15057.4/14811.5	672/704	3.53/3.38	447.6/725.6	138.77/133.65	0/0	0/0	0/0	0/0	0/0

Table 6.6 A summary of performance on the 'B' problems. Results are given in the format $< unseeded > / < seeded >$. These results are based on 10 runs, the Size is the number of unique solutions created over 10 runs. The solution characteristics are the lowest value observed during 10 runs

Problem	Size	Avg. fitness	Fixed cost	Cost per delivery	Staff cost	Running cost	Emissions	% cycle deliveries	% cycle dist	Cycles used	Vans used
B-n31-k5-1	4377/**2535**	4481.5/**4403.2**	240/240	**2.18**/2.31	157.8/**153.8**	42.06/**37.15**	0/0	0/0	0/0	0/0	0/0
B-n34-k5-1	5859/**3053**	5504.9/**5361.7**	**224**/272	**2.34**/2.41	**107.2**/183.6	44.21/**42.43**	0/0	0/0	0/0	0/0	0/0
B-n35-k5-1	4834/**2502**	6718.4/**6694.7**	**272**/288	**2.68**/2.69	**195.4**/235.6	57.32/**54.02**	0/0	0/0	0/0	0/0	0/0
B-n38-k6-1	4210/**2878**	5637.7/**5525.1**	**272**/368	**2.18**/2.28	**184**/215.4	48.63/**45.58**	0/0	0/0	0/0	0/0	0/0
B-n39-k5-1	6228/**3912**	5123.3/**4990.4**	**304**/416	**2.59**/2.99	**169.2**/206.4	39.64/**37.67**	0/0	0/0	0/0	0/0	0/0
B-n41-k6-1	5117/**2814**	6317.7/**6235.9**	**400**/448	2.18/**2.05**	239/**216.4**	58.11/**53.76**	0/0	0/0	0/0	0/0	0/0
B-n43-k6-1	5239/**2748**	5288.3/**5214.6**	384/384	2.55/**2.28**	215.4/**198.6**	47.65/**43.75**	0/0	0/0	0/0	0/0	0/0
B-n44-k7-1	5217/**2612**	6917.4/**6620.7**	**288**/448	**2.4**/2.57	**123.8**/250.6	62.1/**57.85**	0/0	0/0	0/0	0/0	0/0
B-n45-k5-1	5662/**3088**	5605.2/**5433.5**	**416**/464	3.01/**2.91**	**109.0**/242.4	47.14/**42.45**	0/0	0/0	0/0	0/0	0/0
B-n45-k6-1	5145/**2475**	5654.9/**5410.9**	**368**/528	**2.39**/2.62	125.6/**210**	45.3/**45.69**	0/0	0/0	0/0	0/0	0/0
B-n50-k7-1	5515/**2895**	6045.8/**5836.2**	**352**/496	2.61/**2.46**	254.2/**175.4**	52.56/**45.95**	0/0	0/0	0/0	0/0	0/0
B-n50-k8-1	3971/**2222**	8223.2/**8046.3**	**448**/544	2.44/**2.17**	328.6/**271.2**	80.22/**75.59**	0/0	0/0	0/0	0/0	0/0
B-n51-k7-1	4060/**2170**	8412.2/**8293.8**	**448**/548	**2.57**/2.92	402.2/**165**	74.06/**72.08**	0/0	0/0	0/0	0/0	0/0
B-n52-k7-1	5379/**2983**	6871.6/**6740.9**	**352**/512	2.69/**2.66**	**176**/195.4	57.84/**53.26**	0/0	0/0	0/0	0/0	0/0
B-n56-k7-1	6422/**3570**	6902.5/**6649.5**	**464**/512	2.94/**2.72**	219.2/**174.6**	54.71/**51.63**	0/0	0/0	0/0	0/0	0/0
B-n57-k7-1	4350/**2204**	9635.4/**9567**	496/**480**	**2.83**/2.85	**190.4**/229.4	83.06/**80.56**	0/0	0/0	0/0	0/0	0/0
B-n57-k9-1	2978/**1750**	12067.5/**11942.3**	**496**/656	**2.81**/3.24	**378.4**/511.6	119.78/**114.71**	0/0	0/0	0/0	0/0	0/0
B-n63-k10-1	3818/**2107**	11438.7/**11313.5**	**464**/624	**2.4**/2.65	412.2/**369.2**	106.16/**98.4**	0/0	0/0	0/0	0/0	0/0
B-n64-k9-1	5278/**2772**	8094.9/**7823.5**	**576**/736	**2.29**/2.39	297/**205**	65.45/**62.36**	0/0	0/0	0/0	0/0	0/0
B-n66-k9-1	3275/**1832**	11039.5/**10619.8**	**544**/704	**2.75**/3.12	**201.8**/325.8	102.58/**97.89**	0/0	0/0	0/0	0/0	0/0
B-n67-k10-1	4816/**2520**	8896.7/**8539**	**576**/740	2.34/**2.46**	336/**350.6**	75.87/**74.6**	0/0	0/0	0/0	0/0	0/0
B-n68-k9-1	3763/**2111**	11084.9/**10966.3**	**528**/752	**2.8**/3.21	**388.8**/391.0	105.34/**97.66**	0/0	0/0	0/0	0/0	0/0
B-n78-k10-1	3977/**2144**	10817.2/**10526.5**	**704**/992	**2.98**/3.13	465/**297.8**	99.87/**95.09**	0/0	0/0	0/0	0/0	0/0

Table 6.7 A summary of performance on the 'P' problems. Results are given in the format $<$ *unseeded* $>$ / $<$ *seeded* $>$. These results are based on 10 runs, the Size is the number of unique solutions created over 10 runs. The solution characteristics are the lowest value observed during 10 runs

Problem	Size	Avg. fitness	Fixed cost	Cost per delivery	Staff cost	Running cost	Emissions	% cycle deliveries	% cycle dist	Cycles used	Vans used
P-n16-k8-1	8744/**4046**	1353.5/**1260.2**	**96**/160	**1.36**/1.55	**26.8**/31.8	9.82/9.82	0/0	0/0	0/0	0/0	0/0
P-n19-k2-1	7603/**3594**	1741.8/**1575.6**	**112**/160	1.34/1.36	42.2/43.6	12.81/12.81	0/0	0/0	0/0	0/0	0/0
P-n20-k2-1	7965/**3631**	1778.8/**1661.3**	112/144	1.47/**1.42**	46/50.6	12.98/12.98	0/0	0/0	0/0	0/0	0/0
P-n21-k2-1	7490/**3148**	1834.9/**1711.4**	128/128	1.59/**1.43**	55.4/57.6	13.32/**13.14**	0/0	0/0	0/0	0/0	0/0
P-n22-k2-1	7533/**3274**	1944.1/**1777.9**	160/**144**	1.68/**1.47**	46.8/**37.8**	14.13/**13.51**	0/0	0/0	0/0	0/0	0/0
P-n22-k8-1	10124/**6011**	**6681**/6696.8	**128**/256	0.02/0.02	57.0/**40.6**	17.48/17.48	0/0	0/0	0/0	0/0	0/0
P-n23-k8-1	7862/**4679**	1921.3/**1805.3**	**176**/240	**1.6**/1.79	50.6/58	14.12/**13.44**	0/0	0/0	0/0	0/0	0/0
P-n40-k5-1	6982/**3393**	3331.6/**3201**	**208** / 240	1.62/**1.5**	124.6/135.4	26.64/25.65	0/0	0/0	0/0	0/0	0/0
P-n45-k5-1	5679/**2924**	3862.9/**3721**	**256**/304	1.67/**1.54**	79.6/111.6	31.65/**31.09**	0/0	0/0	0/0	0/0	0/0
P-n50-k10-1	5334/**3002**	4280.1/**4009.4**	**400**/532	**1.36**/1.4	101.4/149.0	33.12/**32.4**	0/0	0/0	0/0	0/0	0/0
P-n50-k7-1	5767/**2821**	4211.4/**4012.2**	**448**/660	1.36/1.36	106.2/**94**	32.87/**32.62**	0/0	0/0	0/0	0/0	0/0
P-n50-k8-1	5678/**2851**	4328.7/**4067.6**	**384**/532	1.45/**1.36**	139.4/**83.8**	33.08/**32.74**	0/0	0/0	0/0	0/0	0/0
P-n51-k10-1	5586/**3237**	4227.8/**4071.7**	**448**/548	1.76/**1.96**	229.2/**117.2**	34.32/**34.06**	0/0	0/0	0/0	0/0	0/0
P-n55-k10-1	5422/**2888**	4615.1/**4510.7**	**480**/676	**1.38**/1.48	162.4/**154**	36.07/36.6	0/0	0/0	0/0	0/0	0/0
P-n55-k15-1	5175/**2980**	4676.6/**4432.9**	**480**/628	1.38/1.38	92.0/148.6	35.95/35.95	0/0	0/0	0/0	0/0	0/0
P-n55-k7-1	5100/**2860**	4680.5/**4480.5**	**336**/564	**1.36**/1.5	210.8/**147.6**	35.88/37.24	0/0	0/0	0/0	0/0	0/0
P-n55-k8-1	5086/**2834**	4636/**4496.9**	**416**/676	**1.36**/1.5	125.8/154	35.88/37.63	0/0	0/0	0/0	0/0	0/0
P-n60-k10-1	5322/**2532**	5176.3/**5041.6**	**512**/676	1.39/1.5	148.8/167	40.83/40.96	0/0	0/0	0/0	0/0	0/0
P-n60-k15-1	4855/**2839**	5277.2/**5005.4**	**432**/660	1.41/**1.39**	102.6/162.2	41.06/**40.86**	0/0	0/0	0/0	0/0	0/0
P-n65-k10-1	4604/**2584**	5826.1/**5560.8**	592/656	1.44/**1.42**	157.8/208.4	45.89/**45.87**	0/0	0/0	0/0	0/0	0/0
P-n70-k10-1	4922/**2418**	6259.2/**6031.8**	**612**/768	**1.43**/1.44	176.2/**166.6**	49.54/**49.33**	0/0	0/0	0/0	0/0	0/0
P-n76-k4-1	4654/**2045**	6666.3/**6491.6**	720/**528**	1.53/1.54	228.2/**185**	53.87/**53.33**	0/0	0/0	0/0	0/0	0/0
P-n76-k5-1	4841/**2087**	6678.9/**6415.5**	656/**528**	1.54/1.58	205.4/208.2	53.67/**53.41**	0/0	0/0	0/0	0/0	0/0
P-n101-k4-1	4186/**1694**	8430.5/**8291.3**	**864**/960	1.89/1.94	356.2/**306.2**	73.52/**73.08**	0/0	0/0	0/0	0/0	0/0

Let us assume that the manager has a number of *constraints* (which must be respected) and a number of *aspirations* which should be met if possible. Some of these factors will be known well in advance, others may only come to notice at the last minute:

- **Constraint C1** Staff shortage only 4 bicycles can be operated.
- **Constraint C2** Vehicle shortage only 2 vans can be operated.
- **Constraint C3** Weather emergency, no bicycles can be used.
- **Aspiration A1** Reduce delivery cost.
- **Aspiration A2** Reduce the staff cost and running costs.
- **Aspiration A3** Make maximum use of cycle couriers for publicity purposes.
- **Aspiration A4** Reduce fixed vehicle costs as much as possible.

On any particular day, the manager may be faced with any combination of the above (Note that *c1* and *c3* are mutually exclusive), in some cases with very little warning. To take an example at random; suppose *C2*, *A2* and *A3* apply on one particular day. Taking the B-n67-K10 problem instance at random with 1 h windows, 2151 solutions are provided to the manager (Fig. 6.3a).

We can use the parallel coordinates tool to apply *C2* to the archive (Fig. 6.3b) restricting the number of vans to 2 or less only leaves a small number of solutions. If we then consider *A2*, by eliminating solutions with a high staff or running costs we are left with one solution (Fig. 6.3c) conveniently this makes maximum use of bicycles.

Our manager can use an interactive Parallel Coordinates tool to experimentally apply the constraints and aspirations that apply on that day to the map in order to find a solution which represents the best match.

Let us now select three sets of constraints/aspirations and search the solutions provided for each problem instance to see if they can be met from within the solutions provided. Table 6.8 shows the effect of applying 3 possible scenarios:

- *A2, A3, C2*: Reduce delivery cost. Make maximum use of cycle couriers for publicity purposes. Vehicle shortage only 2 vans can be operated.
- *A1, A2*: Reduce delivery cost. Reduce the staff cost and running costs.
- *A1, A2, A3* Reduce delivery cost. Reduce the staff cost and running costs. Make maximum use of cycle couriers for publicity purposes.

We apply these filters to each of the 10 sets of solutions produced and record the average number of solutions available after the filters have been applied and the lowest fitness found for the seeded and unseeded solutions. The biggest trend noted is that for the majority of problems the set of solutions produced by the unseeded algorithm does not contain sufficient diversity to leave any solutions after the filters have been applied.

In order to gain some idea of the flexibility of MAP-Elites, we can simulate the supermarket managers' scenario. We can generate every possible combination of constraints and aspirations (up to three items in length) and apply them to our problem instances, recording the choice available to the manager. Table 6.9 shows the results of this experiment. The scenarios are applied to the results obtained from

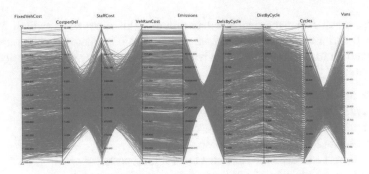

(a) The archive with no filters constraints or aspirations applied

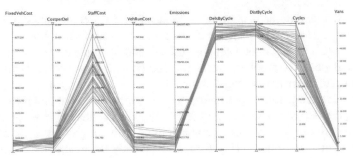

(b) Applying constraint *C2*.

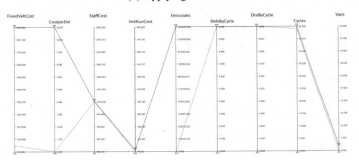

(c) Applying constraints *C2* and *A2*.

Fig. 6.3 The archive of Elite solutions generated for problem B-n67-k10 with 1 h time windows, showing how the constraint *C2* (Vehicle shortage only 2 vans can be operated) and the aspiration *A2* (Reduce the staff cost and running costs) can be applied using an interactive parallel coordinates plot

Table 6.8 The application of 3 scenarios (A2 A3 C2, A1 A2 and A1 A2 A3) applied to the output from each problem in the test instance

| | A2 A3 C2 | | | | A1 A2 | | | | A1 A2 A3 | | | |
| | Seeded | | Not seeded | | Seeded | | Not seeded | | Seeded | | Not seeded | |
	Avg. choice	Best fitness	Avg. choice	Best fitness	Avg. choice	Best fitness	Avg. choice	Best fitness	Avg. choice	Best fitness	Avg. choice	Best fitness
A-n32-k5-1	2.8	4557.56	0	–	6.2	4557.56	0	–	5.2	4557.56	0	–
A-n33-k5-1	10.9	3517.88	0	–	35.7	3437.73	0	–	29.1	3517.88	0	–
A-n33-k6-1	14.9	3463.74	0	–	59.5	3192.53	0	–	40.2	3365.62	0	–
A-n34-k5-1	6.6	4082.05	0	–	23.4	3996.35	0	–	18.4	4082.05	0	–
A-n36-k5-1	1.2	5568.13	0	–	4.1	5210.67	0	–	3.1	5210.67	0	–
A-n37-k5-1	9.7	3475.67	0	–	36.9	3475.67	0	–	28.7	3475.67	0	–
A-n37-k6-1	0.1	6385.61	0	–	1.2	5065.17	0	–	0.5	5269.04	0	–
A-n38-k5-1	13.7	3940.48	0	–	54.6	3940.48	0	–	42.3	3940.48	0	–
A-n39-k5-1	8.4	4962.4	0	–	28.4	4828.48	0	–	23.3	4900.19	0	–
A-n39-k6-1	4.4	4884.21	0	–	25	4680.73	0	–	18.1	4884.21	0	–
A-n44-k7-1	3.3	6032.95	0	–	18.3	5756.86	0	–	14.1	5888.43	0	–
A-n45-k6-1	2.8	5674.13	0	–	21.6	5347.12	0	–	16.8	5674.13	0	–
A-n45-k7-1	0	–	0	–	0.7	7266.72	0	–	0.3	7421.72	0	–
A-n46-k7-1	7.7	5353.54	0	–	36.7	5353.54	0	–	30.2	5353.54	0	–
A-n48-k7-1	1.2	7812.44	0	–	7.2	7372.43	0	–	5.8	7574.33	0	–
A-n53-k7-1	2.9	6922.8	0	–	24.7	6667.13	0	–	21.5	6716.3	0	–
A-n54-k7-1	2.5	8964.85	0	–	6.7	8188.68	0	–	5.9	8438.33	0	–
A-n55-k9-1	3.7	6265.95	0	–	34.5	6245.29	0	–	28.1	6245.29	0	–
A-n60-k9-1	1	9674	0	–	1.1	9674	0	–	1	9674	0	–
A-n61-k9-1	3.7	6389.11	0	–	41.3	6217.04	0	–	32	6217.04	0	–
A-n62-k8-1	0	–	0	–	0.3	10185.42	0	–	0.1	10474.28	0	–
A-n63-k9-1	1.4	11718.28	0	–	10.5	11653.76	0	–	8.1	11653.76	0	–
A-n63-k10-1	0	–	0	–	4.5	7950.37	0	–	3.2	7950.37	0	–
A-n64-k9-1	0	–	0	–	1.1	10309.06	0	–	0.9	10309.06	0	–
A-n65-k9-1	0	–	0	–	6.5	8228.03	0	–	5.5	8324.64	0	–
A-n69-k9-1	2.1	7686.67	0	–	32.3	7535.31	0	–	26.7	7686.67	0	–
A-n80-k10-1	0	–	0	–	0.5	14016.42	0	–	0.2	14590.35	0	–

Table 6.8 (continued)

| | A2 A3 C2 | | | | A1 A2 | | | | A1 A2 A3 | | | |
| | Seeded | | Not seeded | | Seeded | | Not seeded | | Seeded | | Not seeded | |
	Avg. choice	Best fitness	Avg. choice	Best fitness	Avg. choice	Best fitness	Avg. choice	Best fitness	Avg. choice	Best fitness	Avg. choice	Best fitness
B-n31-k5-1	7	4083.21	0	–	20.1	3553.28	0	–	13.7	4003.98	0	–
B-n34-k5-1	6	4892.17	0	–	25.6	4550.72	0	–	16.1	4892.17	0	–
B-n35-k5-1	1.7	6403.57	0	–	4	6403.57	0	–	3.7	6403.57	0	–
B-n38-k6-1	2.3	5080.32	0	–	9.2	4942.86	0	–	7.1	4997.06	0	–
B-n39-k5-1	8.4	4324.06	0	–	39.5	4039.74	0	–	30.9	4040.83	0	–
B-n41-k6-1	8.1	5868.31	0	–	30.1	5378.97	0	–	25	5821.71	0	–
B-n43-k6-1	1.8	4900.84	0	–	10.7	4854.66	0	–	8.6	4900.84	0	–
B-n44-k7-1	0.1	6278.85	0	–	5.7	5765.35	0	–	4.2	6133.11	0	–
B-n45-k5-1	6.4	4834.58	0	–	33.1	4690.78	0	–	26.8	4690.78	0	–
B-n45-k6-1	3.8	4726.08	0	–	24.8	4667.88	0	–	20.7	4726.08	0	–
B-n50-k7-1	4.1	5294.63	0	–	25.6	5294.63	0	–	19.8	5294.63	0	–
B-n50-k8-1	1	9576.84	0	–	1.2	9121.97	0	–	1.2	9121.97	0	–
B-n51-k7-1	0	–	0	–	0	–	0	–	0	–	0	–
B-n52-k7-1	3.3	6277.77	0	–	22.2	5969.84	0	–	17.4	6189.58	0	–
B-n56-k7-1	3.7	5783.61	0	–	29.6	5537.03	0	–	24.3	5783.61	0	–
B-n57-k7-1	1.8	9552.74	0	–	14.5	8953.22	0	–	11.7	9372.89	0	–
B-n57-k9-1	0	–	0	–	0.8	12934.73	0	–	0.3	13504.34	0	–
B-n63-k10-1	1.2	11215.65	0	–	8.8	10540.8	0	–	6.3	10540.8	0	–
B-n64-k9-1	2	6446.75	0	–	28.1	6324.91	0	–	21.9	6446.75	0	–
B-n66-k9-1	0	–	0	–	0	–	0	–	0	–	0	–
B-n67-k10-1	2.2	7573.87	0	–	28.3	7304.98	0	–	24	7520.49	0	–
B-n68-k9-1	0	–	0	–	0	–	0	–	0	–	0	–
B-n78-k10-1	0	–	0	–	6.2	10146.23	0	–	5.2	10146.23	0	–

Table 6.8 (continued)

| | A2 A3 C2 | | | | A1 A2 | | | | A1 A2 A3 | | | |
| | Seeded | | Not seeded | | Seeded | | Not seeded | | Seeded | | Not seeded | |
	Avg. choice	Best fitness	Avg. choice	Best fitness	Avg. choice	Best fitness	Avg. choice	Best fitness	Avg. choice	Best fitness	Avg. choice	Best fitness
P-n16-k8-1	44.1	931.71	40	982.84	95.2	914.09	2.1	1389.54	54.1	931.71	2.1	1389.54
P-n19-k2-1	45.3	1230.66	15.2	1318.06	105.4	1170.92	0	–	62.8	1230.66	0	–
P-n20-k2-1	39.5	1238.04	47.4	1315.37	77	1238.04	2.7	1321.62	54	1238.04	2.7	1321.62
P-n21-k2-1	48.1	1314.78	61.4	1364.29	117	1255.16	0	–	77.9	1314.78	0	–
P-n22-k2-1	42.8	1351.52	13.8	1416.91	98.6	1264.95	0	–	69.9	1351.52	0	–
P-n22-k8-1	65.9	1748.26	0	–	331	1748.26	0	–	117.3	1748.26	0	–
P-n23-k8-1	38.3	1298.94	0.1	1498.79	72	1298.94	0	–	53.7	1298.94	0	–
P-n40-k5-1	22.6	2584.94	7	2754.11	97.2	2563.06	0	–	75.5	2584.94	0	–
P-n45-k5-1	14.8	3081.7	2.1	3237.52	64.1	3062.73	0	–	55.3	3062.73	0	–
P-n50-k7-1	11.3	3262.63	9	3287.67	67	3254.42	0	–	57.4	3262.63	0	–
P-n50-k8-1	8.8	3259.18	9.7	3303.32	59.8	3244.49	0	–	52.4	3244.49	0	–
P-n50-k10-1	11.4	3240.5	6	3312.4	61.5	3240.5	0	–	52.4	3240.5	0	–
P-n51-k10-1	14.3	3418.84	7.5	3547.88	66.3	3393	0	–	57.3	3393	0	–
P-n55-k15-1	5	3595.08	9.7	3595.08	41.9	3595.08	0	–	36.4	3595.08	0	–
P-n55-k7-1	12.3	3594.12	7.7	3588.11	71.9	3594.12	0	–	62.9	3594.12	0	–
P-n55-k8-1	9.9	3592.79	5.6	3588.11	72.2	3592.12	0	–	63.3	3592.12	0	–
P-n55-k10-1	8.2	3545.17	1	3607.54	52.5	3545.17	0	–	45.8	3545.17	0	–
P-n60-k10-1	6.5	4096.78	7	4083.49	58.8	4064.03	0	–	51.9	4096.78	0	–
P-n60-k15-1	6.4	4086.66	7	4106.1	52.4	4048.13	0	–	45.6	4048.13	0	–
P-n65-k10-1	5.4	4524.67	6	4589.34	63.2	4524.67	0	–	54.3	4524.67	0	–
P-n70-k10-1	4.9	4933.96	4.5	4954.74	56.1	4933.96	0	–	49.3	4933.96	0	–
P-n76-k4-1	6.4	5333.57	1	5387.84	58.2	5300.08	0	–	55.6	5300.08	0	–
P-n76-k5-1	7.1	5341.27	1	5357.09	80.1	5262.49	0	–	76	5262.49	0	–
P-n101-k4-1	5.9	7329.24	1	7382.73	51.8	7329.24	0	–	48.6	7329.24	0	–

Table 6.9 The effects of generating all scenarios containing up to three constraints/aspirations and applying them to each problem instance. The number of solutions left for the user to browse is recorded (0 implies that no solutions matched the scenario). Wins refers to the number of problem instances where either seeded or unseeded contained the solution with the lowest fitness after the scenario had been applied

Scenario	A problems				B problems				P Problems			
	Avg.Sols. seeded	Avg. Sols. unseeded	Seeded wins	Unseeded wins	Avg.Sols. seeded	Avg. Sols. unseeded	Seeded wins	Unseeded wins	Avg.Sols. seeded	Avg.Sols. unseeded	Seeded wins	Unseeded wins
A1	1576.69	0	27	0	1583.36	0.19	23	0	1858.14	0.51	21	0
A1 A2	19.39	0	27	0	19.3	0.05	20	0	82.13	0.12	22	0
A1 A2 A3	15.16	0	27	0	14.29	0.05	20	0	59.57	0.12	22	0
A1 A2 A4	14.78	0	27	0	14.78	0.05	20	0	58.25	0.12	22	0
A1 A2 C1	0	0	0	0	0	0	0	0	0.14	0	1	0
A1 A2 C2	3.94	0	21	0	5.65	0.05	18	0	23.88	0.12	22	0
A1 A2 C3	0	0	0	0	0	0	0	0	0	0	0	0
A1 A3	298.58	0	27	0	297.13	0.19	23	0	322.09	0.51	21	0
A1 A3 A4	248.62	0	27	0	246.09	0.19	23	0	271.79	0.51	21	0
A1 A3 C1	0	0	0	0	0	0	0	0	0	0	0	0
A1 A3 C2	67.51	0	27	0	66.53	0.19	23	0	94.8	0.51	21	0
A1 A3 C3	0	0	0	0	0	0	0	0	0	0	0	0
A1 A4	369.72	0	27	0	368.5	0.19	23	0	394.12	0.51	21	0
A1 A4 C1	0	0	0	0	0.85	0	1	0	1.73	0	7	0
A1 A4 C2	73.14	0	27	0	77.92	0.19	23	0	119.09	0.51	21	0
A1 A4 C3	0	0	0	0	0	0	0	0	0	0	0	0

(continued)

Table 6.9 (continued)

Scenario	A problems				B problems				P Problems			
	Avg.Sols. seeded	Avg. Sols. unseeded	Seeded wins	Unseeded wins	Avg.Sols. seeded	Avg. Sols. unseeded	Seeded wins	Unseeded wins	Avg.Sols. seeded	Avg. Sols. unseeded	Seeded wins	Unseeded wins
A1 C1	137.75	0	27	0	172.92	0	24	0	316.62	0	24	0
A1 C1 C2	0	0	0	0	0.14	0	1	0	0.14	0	1	0
A1 C1 C3	37.35	0	27	0	40.23	0	24	0	43.89	0	24	0
A1 C2	73.14	0	27	0	77.92	0.19	23	0	119.09	0.51	21	0
A1 C2 C3	0	0	0	0	0	0	0	0	0	0	0	0
A1 C3	37.35	0	27	0	40.23	0	24	0	43.89	0	24	0
A2	19.39	0	27	0	19.3	0.26	20	0	82.13	6.7	0	0
A2 A3	15.16	0	27	0	14.29	0.26	20	0	59.57	3.61	0	0
A2 A3 A4	12.93	0	27	0	12.17	0.1	20	0	50.64	1.12	2	0
A2 A3 C1	0	0	0	0	0	0	0	0	0	0	0	0

(continued)

Table 6.9 (continued)

Scenario	A problems				B problems				P Problems			
	Avg.Sols. seeded	Avg. Sols. unseeded	Seeded wins	Unseeded wins	Avg.Sols. seeded	Avg. Sols. unseeded	Seeded wins	Unseeded wins	Avg.Sols. seeded	Avg. Sols. unseeded	Seeded wins	Unseeded wins
A2 A3 C2	3.89	0	21	0	4.54	0.26	18	0	20.22	2.7	1	0
A2 A3 C3	0	0	0	0	0	0	0	0	0	0	0	0
A2 A4	14.78	0	27	0	14.78	0.1	20	0	58.25	1.13	2	0
A2 A4 C1	0	0	0	0	0	0	0	0	0	0	0	0
A2 A4 C2	3.94	0	21	0	5.65	0.1	18	0	23.88	1.13	2	0
A2 A4 C3	0	0	0	0	0	0	0	0	0	0	0	0
A2 C1	0	0	0	0	0	0	0	0	0.14	0.01	0	0
A2 C1 C2	0	0	0	0	0	0	0	0	0	0	0	0
A2 C1 C3	0	0	0	0	0	0	0	0	0	0	0	0
A2 C2	3.94	0	21	0	5.65	0.26	18	0	23.88	2.72	1	0
A2 C2 C3	0	0	0	0	0	0	0	0	0	0	0	0

(continued)

Table 6.9 (continued)

Scenario	A problems				B problems				P Problems			
	Avg.Sols. seeded	Avg. Sols. unseeded	Seeded wins	Unseeded wins	Avg.Sols. seeded	Avg. Sols. unseeded	Seeded wins	Unseeded wins	Avg.Sols. seeded	Avg. Sols. unseeded	Seeded wins	Unseeded wins
A2 C3	0	0	0	0	0	0	0	0	0	0	0	0
A3	298.58	438.91	0	0	297.13	432.99	0	0	322.09	587.89	0	0
A3 A4	248.62	62.1	0	0	246.09	73.2	0	0	271.79	114.57	0	0
A3 A4 C1	0	0	0	0	0	0	0	1	0	0	0	1
A3 A4 C2	67.51	54.02	0	0	66.53	54.13	0	0	94.8	87.63	0	0
A3 A4 C3	0	0	0	0	0	0	0	0	0	0	0	0
A3 C1	0	0	0	0	0	0.02	0	1	0	0.02	0	1
A3 C1 C2	0	0	0	0	0	0.01	0	1	0	0.01	0	1
A3 C1 C3	0	0	0	0	0	0	0	0	0	0	0	0
A3 C2	67.51	98.64	0	0	66.53	99.07	0	0	94.8	196.76	0	0
A3 C2 C3	0	0	0	0	0	0	0	0	0	0	0	0

(continued)

Table 6.9 (continued)

Scenario	A problems				B problems				P Problems			
	Avg.Sols. seeded	Avg. Sols. unseeded	Seeded wins	Unseeded wins	Avg.Sols. seeded	Avg. Sols. unseeded	Seeded wins	Unseeded wins	Avg.Sols. seeded	Avg. Sols. unseeded	Seeded wins	Unseeded wins
A3 C3	0	0	0	0	0	0	0	0	0	0	0	0
A4	369.72	65.98	0	0	368.5	81.17	0	0	394.12	127.43	0	0
A4 C1	0	0	0	0	0.85	0	0	0	1.73	0.01	4	0
A4 C1 C2	0	0	0	0	0.14	0	0	0	0.14	0.01	0	2
A4 C1 C3	0	0	0	0	0	0	0	0	0	0	0	0
A4 C2	73.14	57.03	0	0	77.92	59.76	0	0	119.09	99.6	0	0
A4 C2 C3	0	0	0	0	0	0	0	0	0	0	0	0
A4 C3	0	0	0	0	0	0	0	0	0	0	0	0
C1	137.75	14.4	0	0	172.92	25.6	0	0	316.62	51.26	0	0
C1 C2	0	0	0	0	0.14	0.09	0	0	0.14	0.1	0	3
C1 C2 C3	0	0	0	0	0	0	0	0	0	0	0	0
C1 C3	37.35	3.38	0	0	40.23	3.89	0	0	43.89	5.32	0	0
C2	73.14	104.59	0	0	77.92	122.55	0	0	119.09	256.54	0	0
C2 C3	0	0	0	0	0	0	0	0	0	0	0	0
C3	37.35	3.38	0	0	40.23	3.89	0	0	43.89	5.32	0	0

the seeded and unseeded versions of the algorithm. In the majority of cases, the seeded version of the algorithm provides sufficient diversity of solutions in order to find some solutions that match the criterion. There are a few criterion sets (e.g. A4, C3) where no solutions are found. In this case, criterion would be

Reduce fixed vehicle costs as much as possible. (A4)

Weather emergency, no bicycles can be used. (C3)

If we apply wider domain knowledge, we can assume that in this case C3 would take priority over A4 and it is unlikely that would be used in combination. Once again we note that the seeded algorithm provides sets of answers of a greater diversity.

6.5 Conclusions

The scenario discussed in this chapter have two fundamental properties:

- The solution has many characteristics which are of interest to the user.
- The user has the expertise and experience to determine which solution characteristics are of greater importance when choosing a solution.

We are, in a sense, taking a more holistic view of optimisation. We accept that although algorithms can be powerful tools they are just a tool and perform best when applied by a skilled user. The usefulness of the algorithm is in acting as a solution filter, starting with an incomprehensibly large search space and reducing it to a far smaller set of *elite* solutions that are of potential interest to the user. The set of *elite* solutions represents a conflict; on the one hand, we wish it to be large enough to encompass a range of solutions that illustrate the dimensions of the solution space, whilst on the other hand a large set of solutions (e.g. several thousand) may be as incomprehensible to the user as the initial solution space.

We seek to solve this problem by using interactive Parallel Coordinates plots. With training an expert user can use the PC plot to explore the characteristics of the elite solutions in order to find one that matches their requirements.

Our approach may be summarised as

1. Execute *MAP-Elites* to produce a set of elites.
2. Create a parallel coordinates plot of the elites.
3. Allow the user to interact with the parallel coordinates plot to explore the range of solutions available.

Step one may require some significant CPU time, but once completed, steps 2 and 3 can be carried out interactively with the user in real time. In a real-world scenario, where the production of a solution is time constrained we don't want to repeat step one if at all possible. The user can apply their domain expertise at stage 3 to find a solution that best meets their current requirements. Given that some requirements may alter with a very short timescale (e.g. staff or vehicle availability or weather related issues), the ability to select a different solution by "browsing" the elite set is far more practical than having to run the solver once more and await a new solution.

References

Augerat, P. 1995. *Approche polyédrale du probléme de tournées de véhicules. (Polyhedral approach of the vehicle routing problem)*, Ph.D. Thesis. France: Grenoble Institute of Technology. https://tel.archives-ouvertes.fr/tel-00005026.

Augerat, P. 2014a. VRP-REP: Augerat 1995 Set A. http://www.vrp-rep.org/datasets/item/2014-0000.html.

Augerat, P. 2014b. VRP-REP: Augerat 1995 Set B. http://www.vrp-rep.org/datasets/item/2014-0001.html.

Augerat, P. 2014c. VRP-REP: Augerat 1995 Set P. http://www.vrp-rep.org/datasets/item/2014-0009.html.

Holland, J. H. 1975. *Adaptation in Natural and Artificial Systems*. University of Michigan Press.

Inselberg, A. 2009. *Parallel Coordinates: Visual Multidimensional Geometry and Its Applications*. Advanced Series in Agricultural Sciences. New York: Springer.

Mouret, J.-B., Clune, J. 2015, Illuminating Search Spaces by Mapping Elites, (1504.04909). https://arxiv.org/abs/1504.04909.

Part II
Data and Algorithms

This section discusses the underlying data structures and algorithms required in order to apply the problems discussed in Part 1 to real-world applications.

We discuss the nature of maps and geospatial data and examine the relationship between traditional cartographic techniques and current technologies. We examine techniques for pathfinding through road networks, examining how geospatial data may be converted into data structures representing road networks. The use of algorithms such as A* and Dijkstra's pathfinding algorithm to find paths through road networks is examined.

Emphasis is placed on the practicalities of building software systems that use readily available data from OpenStreetMap. The reader is taken through the process of constructing a simple, but functional, routing engine.

Chapter 7
GeoSpatial Data

Abstract Graphical Information Systems utilise projection and coordinate systems derived from traditional cartography. In order to better understand these concepts undertake a review of the history of cartography and map making. We also describe the fundamental principles of map projection as well and the latitude/longitude coordinate system. The relationship between cartography and computer graphics is explored and the use of raster and vector formats is discussed. A brief review of graph theory is presented, and use of graphs to represent transportation networks (e.g. road networks) is described.

7.1 A Brief History of Mapping

Traditionally, geospatial data has been recorded in the visual form of maps, created by *cartographers*. The earliest attempts to record the world around us in map form date back to ancient times and appear to have centred on mapping the position of stars in the sky rather than earthbound features Whitehouse (2000). For an in-depth history of cartography, the reader is directed to Wilford (1981).

The first documents that we might recognise as a map date the middle-ages, such as the Hereford Mappa Mundi (Fig. 7.1). As explorers journeyed over more of the planet, the maps created by *cartographers* contained more information and increased in accuracy.

The acceptance that the Earth is a globe and not flat posed a problem for cartographers as the surface of a globe cannot be easily translated into 2 dimensions. It was Gerardus Mercator (1512–1594) Crane (2004) who devised a scheme known as *Mercator's Projection* that allowed reasonably accurate maps of the surface of the Earth to be drawn in 2 dimensions. Mercator's 1569 map of the world (Fig. 7.2) presents a form that is more familiar to us.

As surveying techniques developed so map making became more accurate and to scale. One of the major forces driving developments in cartography and surveying were the requirements of military commanders who needed accurate maps of terrain in order to formulate strategies. In the United Kingdom, military commanders required an accurate map of the Highlands of Scotland in order to deal with the

© Springer Nature Switzerland AG 2022

N. Urquhart, *Nature Inspired Optimisation for Delivery Problems*,
Natural Computing Series, https://doi.org/10.1007/978-3-030-98108-2_7

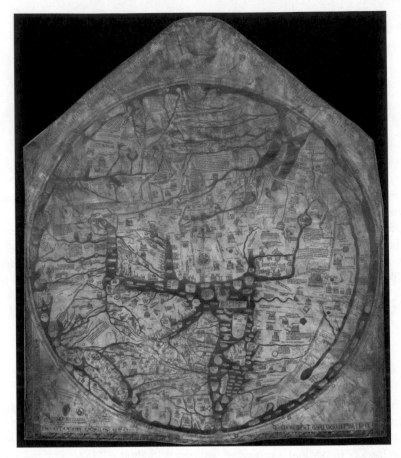

Fig. 7.1 The 13th century Hereford Mappa Mundi. unesco.org.uk [Public domain PD-US], *Source* UNESCO, Author: Unknown https://en.wikipedia.org/wiki/Mappa_mundi#/media/File:Hereford-Karte.jpg

aftermath of the Jacobite uprising of 1745. The British army undertook an accurate survey of the Highlands commencing in 1747, the eventually lead to the formation of the Ordnance Survey organisation Ordnance Survey (2021) who are still producing maps and geographical data to this day.

In 1854, the City of London was ravaged by a cholera outbreak centred on the Soho district Begum (2016); Snow (1855). At that time there were two principle theories as to how cholera spread, one argued that it was airborne (the miasma theory) and the other that it was spread by germs. Whilst investigating the outbreak, physician Jon Snow plotted the cholera cases on a map (see Fig. 7.3). By visualising the outbreak on a map, Snow was able to determine that the cases were clustered around a communal water pump which, upon investigation, was found to be source of the contamination. It has been suggested that Snows' use of maps to plot cholera cases and the subsequent conclusions drawn represent the first significant use of

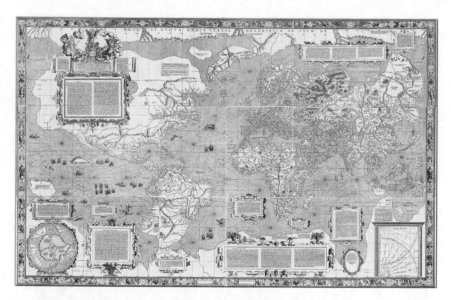

Fig. 7.2 Mercator's 1569 map of the world [Public Domain PD-US] Author: Gerardus Mercator (1569) https://commons.wikimedia.org/wiki/File:Mercator_1569.png

maps to visualise data allowing conclusions to be drawn that would not have been obvious by other means.

The development of printing techniques allowed for the widespread production and distribution of maps and atlases during the twentieth century. The use of maps by individuals for navigation and leisure use became widespread in many countries. The development of computing hardware and software in the second half of the twentieth century lead to the development of the *Geographical Information System* (GIS) Goodchild (2018); Tomlinson and Inventory (1967).

A Graphical Information System uses computers to process, store and visualise geographical data. As data storage and graphical capabilities increased GIS were developed to allow users to browse maps on screen and have data overlaid onto the maps. Other common GIS functions include the ability to search data on a geospatial basis, e.g. find me the closest restaurant to a specific location. Other useful functions of GIS include finding routes between locations in order to generate driving directions. Until the advent of the world wide web, GIS remained expensive and specialised. Early users of GIS included utility organisations to keep track of pipes and cables or city planning departments to keep track of land use.

The first instance of maps being made accessible on line was the Xerox Parc Map Viewer in 1993 Putz (1994) which lead to the development of web mapping platforms such as Google Maps Google Maps (2021). This has diminished the role of the specialised GIS package, as many common GIS functions are now provided by publicly available web mapping services. The principle behind web mapping is to have map data stored on a server and allow client web browsers to display maps on demand. Initially web mapping services were *tile based*, the map being divided into

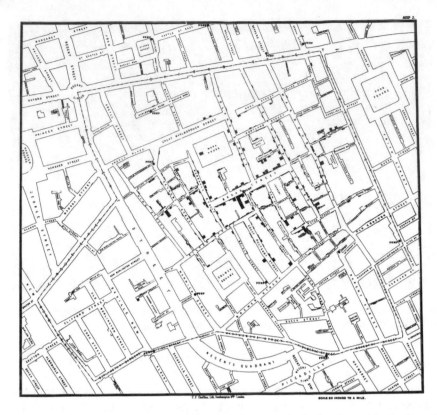

Fig. 7.3 The 1854 map by John Snow showing cholera cases in the London epidemic, from the book "On the Mode of Communication of Cholera" by John Snow, originally published in 1854 by C.F. Cheffins, Lith, Southhampton Buildings, London, England. [Public Domain PD-US] https:// commons.wikimedia.org/wiki/File:Snow-cholera-map-1.jpg

tiles which were stored as images, being sent to the browser on demand. Tile-based services are largely being superseded by vector-based web mapping (see Sect. 7.2).

Web mapping services have become increasingly interactive, allowing users to plan journeys, search for points of interest, mix mapping data and satellite imagery as well as uploading data which may be overlaid onto the map. Most mainstream providers of web mapping services provide one or more Application Program Interfaces (APIs) to allow software developers to integrate the mapping service into their software.

Map publishing has traditionally been an area where cartographic organisations have used copyright laws to control access to their products and data. Traditional GIS packages were expensive due to the need to license copyrighted data. Web Mapping services have usually been free for individuals to use, but API access is normally restricted and requires a license for commercial use. A recent development has been the rise of *collaborative mapping*, where users supply data on a voluntary basis to build a map. Collaborative maps such as Open Streetmap (OSM) Haklay and Weber (2008) are normally licensed under open-source agreements allowing them to be used free of charge. OSM in particular is able to offer a data set that compares favourably with commercial offerings in many geographical areas.

Within this book, we will make use of Open Streetmap data OpenStreetMap contributors (2021) and related open-source software. OSM allows unrestricted use of its mapping data and avoids any issues around licensing. It also allows access to raw data in a manner that commercial web mapping services cannot offer.

7.2 Graphics: Raster Versus Vector

Maps are fundamentally a visual medium, presenting a representation of a geographical district to the viewer. In recent times GIS have allowed the creation of custom maps, with the user selecting the area, the scale and possibly providing additional data to be overlaid into the map. The development of GIS and web mapping has been facilitated by technological developments in computer graphics and imaging. Image data may be classified into one of two principle types:

- **Raster**: Raster formats encode a picture as a series of bits (sometimes referred to as a *bitmap*).
- **Vector**: The image is encoded as set of points which are connected by lines.

Raster images are easily displayed, typically they can be loaded into the memory of a video card with little additional processing. Some more recent formats such as JPEG ISO/ IEC (1994) feature compression and must decompressed before display. A vector image may be thought of as a set of instructions that describe how the image should be drawn. The drawing of an image for display using vector graphics is known as *rendering* the image.

Both formats have their advantages, raster images are quick to display and are easily generated from cameras and document scanners. The size of a raster image can be an issue as can the difficulties in processing the data within the image. For instance, a vector image of a map is simply a bitmap comprising a binary string; they is nothing within the format that describes the streets or other features on the map represented by the bitmap. Image recognition techniques can allow features to be extracted from raster images, although this can be a time consuming process.

In many respects vector-based graphics are a good choice for representing map data;

- Vectors may be scaled upon rendering allowing the resulting map to be displayed at a number of scales.
- Items within the maps can be eliminated from the rendering at certain scales or at the behest of the user.
- The vector data can contain additional, non-rendered data items to allow the data to be used for additional services such as routing or searching.

A widely used example of a vector based format is the .OSM format used by Open Streetmap OpenStreetMap contributors (2021) (see Fig. 7.4). In this case, the file format is based on XML standards allowing data to be structured and tagged.

```
<way id="4237322" visible="true" version="16" changeset="46283903"
timestamp="2017-02-21T19:07:21Z" user="eric_" uid="1798469">
<nd ref="606364"/>
<nd ref="25298217"/>
<nd ref="25298216"/>
<nd ref="25298215"/>
<nd ref="25298214"/>
<nd ref="626112"/>
<nd ref="626139"/>
<nd ref="626134"/>
<nd ref="626129"/>
<nd ref="366053712"/>
<nd ref="4464837737"/>
<nd ref="366053705"/>
<nd ref="626119"/>
<tag k="highway" v="pedestrian"/>
<tag k="name" v="Rose Street"/>
<tag k="oneway" v="yes"/>
<tag k="source" v="Bing;survey"/>
<tag k="surface" v="paving_stones"/>
</way>
```

Fig. 7.4 An example of data contained within a .OSM vector file, this extract relates to Rose Street in Edinburgh which appears in Fig. 7.7a

Web mapping systems typically use a vector format for storing their maps; the raw vector data can then be rendered into raster-based images, known as *tiles*. Each tile comprises a section of the map and is stored in a cache on the server. As a user views the map within their browser, tiles are sent from the server for display. Tiles may also be cached at different scales in order to facilitate zooming in and out of the map. Raster-based tiles have the advantage that the rendering may be carried out off-line within the server on a regular basis as the underlying vector data is updated. The principle disadvantage is the amount of raster image data that has to be transferred from the server to the client. This can be particularly obvious when the user is scrolling through a map and has to pause for tiles to be received.

A recent development is the use of vector-based tiles Vladimir Agafonkin et al. (2019). This has the advantage of requiring less data to be transmitted to the browser. The current generation of browsers has sufficient capabilities to render the tiles within the browser.

7.3   Coordinates Systems

A geographical coordinate system allows the position of any point on the surface of the planet to be accurately recorded, when used in conjunction with maps, it facilitates cross-referencing between maps and reality.

By far the most common system used is that of *latitude, longitude and elevation*. Latitude specifies the distance north or south of the equator, the equator being 0°

latitude, the north pole being 90° and the south pole being –90°. Longitude measures the distance to the east or west of the *prime meridian* which runs from the North Pole to the South Pole via Greenwich, UK. To the east of the Prime Meridian longitude is measured in positive degrees up to 180°. To the West of the Prime Meridian longitude is measured as negative from 0 to –180. As the Earth is a sphere longitude –180 and 180 are in effect the same. Elevation measures the height above sea level, and allows a point on the surface of the planet to be fixed in 3 dimensions, in the problems examined in the book, we will only use latitude and longitude without specifying the elevation. The current agreed standard for the Latitude and Longitude is WGS84, which is used by the Global Positioning System and most software systems.

The useful aspect of latitude and longitude is the ability to specify an actual point of the surface of the earth and then to find that same point on a map. The ability to specify and record such points as a pair of numbers revolutionised navigation. Within GIS systems any point may be recorded using a pair of double precision numbers, for instance the author would like to be writing this text at (45.492026, 10.608264) rather than in the depths of a Scottish winter.

For most purposes (certainly the problems in this book) specifying latitude and longitude to 6 decimal places is sufficient. The domain for latitude values is therefore:

$$-90.999999 \text{ to } 90.999999$$

And the equivalent for longitude is

$$-180.999999 \text{ to } 180.999999$$

In the Java language, a *double* will suffice to store a latitude or longitude. To store a point a *java.awt.geom.Point2D.Double* object may be used[1] may be used.

It is necessary to be able to calculate the distance between points, for 2D points on a flat surface the euclidean distance calculation may be used:

$$d(\mathbf{p}, \mathbf{q}) = \sqrt{(q_{lat} - p_{lat})^2 + (q_{lon} - p_{lon})^2}$$

This straight line distance is of no use when working with points on a sphere, where the distance must take into account the curvature of the surface of the sphere. When calculating the distance between two latitude/longitude points the *Haversine* formula Inman (1835) should be used.

$$d(\mathbf{p}, \mathbf{q}) = 2r \arcsin\left(\sqrt{\sin^2\left(\frac{q_{lat} - p_{lat}}{2}\right) + \cos(p_{lat})\cos(q_{lat})\sin^2\left(\frac{q_{lon} - p_{lon}}{2}\right)}\right)$$

where

p_{lat}, p_{lon} is the latitude and longitude of point p in radians.

---

[1] https://docs.oracle.com/javase/7/docs/api/java/awt/geom/Point2D.Double.html.

q_{lat}, q_{lon} is the latitude and longitude of point q in radians.
r is the radius of the sphere (the earth)

The Haversine formula provides a useful means of calculating the direct distance between two points on the surface of the planet, but as we shall see when calculating the distance via or road or other transportation networks the distance is likely to be calculated using values in a *graph*.

Latitude and longitude coordinates are not always practical for humans to use when describing locations—for instance, most adults can remember the number and street name of their home, but would not remember the latitude and longitude of their home. The association of real-world locations specified by a geographical coordinates system with an address or other item of data is known as *geolocation*. The easiest way to geolocate a given location is via the Global Positioning System (GPS) GPS Overview (2021). GPS hardware interprets time signals from a number of satellites, allowing the distance from each satellite to be calculated and thus the position of the GPS receiver to be derived. The fitting of GPS receivers to mobile devices makes them accurately location aware.

For devices such as desktop computers and some laptops which are not equipped with GPS receivers, their network IP addresses can be *geolocated* allowing associated hardware and software to become location aware. For instance, many web browsers make use of geolocation for location-specific advertising (in practice geolocation of IP addresses can be error prone as the location given is often that of the ISP rather than the actual user of the IP). For many of the problems examined in this book, the challenge is to geolocate the addresses of customers or other entities to Latitude and Longitude. Many web mapping services, e.g. Google Maps Google Maps (2021) provide a geolocation API, that will take an address string and return a set of coordinates.

A recent development has been What3Words (2021) which have divided the world into 57×10^{12} 3m by 3m squares, each of which is allocated a name comprising of 3 words. For instance the clock tower of the Palace of Westminster, London (better known as Big Ben) is located at "price.beard.sang". The algorithm used to map from the three words to the coordinates and vice-versa is proprietary, but APIs are available to map from words to coordinates and back.

7.4   Graph Theory and Networks

So far within this book we have calculated distances as the most direct line between two points, using the Euclidean or Haversine formulae. Within the real-world problems discussed in the remainder of this book, most of the distances required will be based on paths through road or other transport networks. Such networks may be represented using a *graph* structure. A graph comprises a set of *nodes* which are linked by a set of *edges* (also known as *nodes* and *links*). An example of a simple graph may be seen in Fig. 7.5, which shows 4 nodes, that are linked by 4 edges.

Fig. 7.5 A graph comprising
of 4 nodes, linked by 4 edges

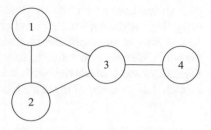

More formally a graph may be described as

$$G = (V, E)$$

where
G is the graph
V is the set of nodes
E is the set of edges

In our example, we see that node 1 has a relationship to nodes 2 and 3, but not
with 4.

At a basic level, graphs allow us to model a collection of nodes and show where
they have a relation to another node (via the presence of an edge) and where no
relationship exists (via the absence of an edge). A major use of graphs in recent
times has been the modelling of networks within social media. The example graph
in Fig. 7.6 shows the relations between 4 individuals. The graph tells us that there is
a relation between Jamie and Katie, for example, and that Jamie and Millie do not
have a relation. We can also see that Katie and Ahmed are both well connected as
they have relations with everyone else within the network. Figure 7.6 introduces the
concept of *weighted edges*. Each of the edges has a numeric weight associated with
it; this could represent the level of activity between the users. For instance, Jamie

Fig. 7.6 A graph showing
relationships in a social
network

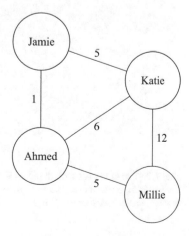

and Ahmed have had relatively little activity, whilst Katie and Millie have a high weighting suggesting a higher level of activity between them.

Our interest in graphs lies in their ability to represent road and transportation networks. Within a road network, each link represents a road and each node a junction. Figure 7.7a and b shows the relationship between a geographically correct map and a graph data structure of the same road network.

(a) A map showing the road network in the centre of Edinburgh.

(b) A graph representation of the centre of Edinburgh

Fig. 7.7 The street network depicted in the upper map can be represented by a graph as shown on the lower map. *Base map and data from OpenStreetMap and OpenStreetMap Foundation* (https://www.openstreetmap.org/copyright)

The graph structure used for such a road network will typically use edge weights to represent road lengths; other attributes that are likely to be associated with edges include

- Street name, e.g. "George Street";
- Road number, e.g. A1;
- Road classification, e.g. (Motorway, Trunk or Minor);
- Speed limit.

Junction nodes may have attributes such as

- Traffic lights (true/false);
- Roundabout (true/false).

If a vector data format is adopted, then the data required to construct road network graph can be included within the vector data. This scheme is used by Open Streetmap within their .OSM file standard OpenStreetMap contributors (2021).

An important aspect of graphs is the orientation of edges. In the examples looked at so far, all the edges have been bidirectional, that is to say that if an edge e connects nodes v_x and v_y then the relation exists v_x to v_y and v_y to v_x. Figure 7.8 shows a directed graph in which the following relations exist (Note that z to x and y to x do not exist.):

- x to y;
- x to z;
- y to z;
- z to y.

The most obvious use of directional edges when representing road networks is to allow for one way streets. Figure 7.9 shows a road network of 4 linked junctions (j_1 to j_4), the road links between j_1 and j_2 and between j_2 and j_4 are bidirectional hence two edges are used in each case. The streets linking j_1 to j_3 and j_3 to j_4 are one way so only one edge is used in each case.

Where junctions are complex and/or have restricted turns multiple nodes and edges may be used to model the junction. Figure 7.10 shows an example of a junction with a restricted turn. Vehicles entering the junction from Napier Street ($S2_1$) arrive at

Fig. 7.8 A directed graph

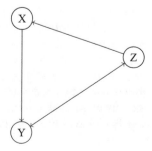

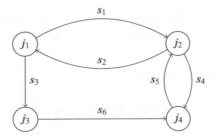

Fig. 7.9 A road network graph based upon a directed graph

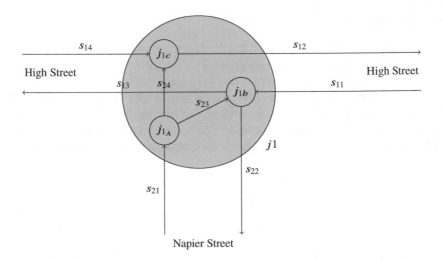

Fig. 7.10 A road junction, with restricted turns represented using a graph

node j_{1a}, they can then proceed to j_{1b} for a left turn onto High Street or j_{1c} for a right turn onto High Street. The arrangement does not allow traffic proceeding along High Street to make a right turn into Napier Street. It also ensures that traffic on High Street cannot perform a "U turn"—e.g. from s_{11} to s_{12}. By using multiple nodes and edges within a junction, complex junctions can be modelled.

There are a number of ways in which we can represent graphs as a data structure. The most obvious scheme is to create Node and Edge classes as suggested in Fig. 7.11. Whilst this design may be considered good practice from an object-oriented perspective, it suffers from scalability issues. The street network for a major city may comprise tens of thousands of nodes and edges. As we will see in Chap. 8 when path finding, it is necessary to traverse the graph structure. The operations required to find whether two nodes are directly linked or to obtain the neighbours (those nodes directly connected to a specific node) of a node will be slower given the number of separate objects that have to be involved in them.

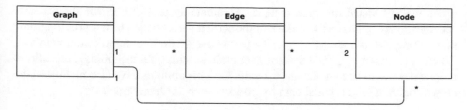

Fig. 7.11 An outline of an object oriented implementation of a graph

Table 7.1 An example of an adjacency matrix, based on the graph shown in Fig. 7.9

	j1	j2	j3	j4
j1	0	1	1	0
j2	1	0	0	1
j3	0	0	0	1
j4	0	1	0	0

An alternative representation is that of the adjacency matrix, each edge is a cell in a 2D matrix, Table 7.1 shows an example. In practice, an array would only have to be *n* by *n* where *n* is the number of nodes in the graph. To determine if two nodes are related, all that is required is an array lookup, if the cell contains a 1 they are connected directly. To find the neighbours of a node, find the row associated with that node, note each column that contains a 1. A useful modification is to have the adjacency matrix holding positive edge lengths to represent a connection and −1 to signify no direct connection.

7.5 Conclusions

This chapter has described the data types and structures commonly used in geographical information systems. Such data forms a necessary part of our problem models when we come to optimise them.

The fundamental concept of creating a map which provides us with a model of the world around us is one that stretches back thousands of years. The cartography process has become digitised and through the medium of web maps, customised maps are now available on demand.

Map data may be stored as raster images or as vector data; a rendering algorithm may be used to create raster images using vector data as an input. Vector data is useful as it can contain additional meta-data connected with features on the map (e.g. the opening times of shops or attractions) or the information required to build graph-based models of road or transport networks.

As well as visual mapping data, a coordinate system that allows the accurate specification of points on the surface of the earth is a necessity. The latitude, longitude system provides the ability to specify points using two double precision numbers. When the latitude longitude system is combined with GPS technology, this allows mobile devices to become location aware. GPS capabilities have also facilitated the crowd sourcing of geospatial data by projects such as Open Streetmap.

References

Begum, F. 2016. Mapping disease: John Snow and Cholera. https://www.rcseng.ac.uk/library-and-publications/library/blog/mapping-disease-john-snow-and-cholera/

Crane, N. 2004. *Mercator: The man who mapped the planet.* Henry Holt and Company: Owl Books.

Goodchild, M. F. 2018. Reimagining the history of GIS, *Annals of GIS* **24**(1), 1–8. Publisher: Taylor & Francis _eprint: https://doi.org/10.1080/19475683.2018.1424737

Google Maps: 2021. https://www.google.com/maps/

GPS Overview: 2021. https://www.gps.gov/systems/gps/

Haklay, M., and P. Weber. 2008. OpenStreetMap: User-generated street maps. *IEEE Pervasive Computing* 7 (4): 12–18.

Inman, J. 1835. *Navigation and nautical astronomy for the use of British seamen.* C. and J. Rivington

ISO/IEC 10918-1:1994: n.d. https://www.iso.org/cms/render/live/en/sites/isoorg/contents/data/standard/01/89/18902.html

OpenStreetMap contributors: 2021. *Planet dump*, retrieved from https://planet.osm.org. Published: https://www.openstreetmap.org

Ordnance Survey. 2021. https://www.ordnancesurvey.co.uk/

Putz, S. 1994. Interactive information services using World-Wide Web hypertext. *Computer Networks and ISDN Systems* **27**(2), 273–280. https://www.sciencedirect.com/science/article/pii/0169755294901414

Snow, J. 1855. *On the Mode of Communication of Cholera.* John Churchill

Tomlinson, R., and Inventory, C. L. 1967. *An Introduction to the Geo-information System of the Canada Land Inventory*, (Department) of Forestry and Rural Development. https://books.google.co.uk/books?id=8fvYSgAACAAJ

Vladimir Agafonkin, John Firebaugh, Eric Fischer, Konstantin Käfer, Charlie Loyd, Tom MacWright, Artem Pavlenko, Dane Springmeyer, and Blake Thompson. 2019. Mapbox Vector Tile Specification, *Technical Report Version 2.1*. https://github.com/mapbox/vector-tile-spec

What3Words. 2021. https://w3w.co

Whitehouse, D. 2000. Ice Age star map discovered. http://news.bbc.co.uk/1/hi/sci/tech/871930.stm

Wilford, J. 1981. *The mapmakers.* A Borzoi Book: Knopf.

Chapter 8
Routing Algorithms

Abstract In order that we may solve real-world instances of the problems discussed earlier (see Sect. 8.1), it becomes necessary to find routes through road networks between deliveries. In this chapter, we examined the means by which geospatial data may be represented and processed in GIS and in particular the use of graphs to represent street networks. We commence this chapter by studying the pioneering work of Dijkstra Dijkstra (1959) in routing and examine a range of algorithms including A* (pronounced A-Star) Hart et al. (1968), Highway Hierarchies Sanders and Schultes (2006) and Contraction Hierarchies Geisberger et al. (2008) all of which may be used to find routes through graphs. We discuss the construction of a simple routing engine, which constructs street graphs based on OpenStreetMap data. We compare Dijkstra and A* and demonstrate the usefulness of hierarchical techniques.

8.1 Routing Algorithms

As discussed previously (Sect. 7.4), a graph data structure is an excellent means of representing a road network. If we can find a route[1] between two nodes, then we can calculate the distance represented by the route; for many of our problems, it is this distance value which is of interest to us.

A number of routing algorithms exist for finding routes between nodes within a graph; in this section, we shall endeavour to compare several of the most fundamental ones.

Routing algorithms require access to the underlying geospatial data. This underlying data takes the form of a graph that represents the underlying street or transportation network. When describing the routing algorithms, we will assume that the underlying data structure will support the following methods:

- *linked(Node n1, Node n2)*
  returns True if n1 and n2 are linked by a Way, else False.

---

[1] A note on terminologies; we will use the term routing to describe the process of finding a route through a graph. Some authors use the terms path finding and paths.

© Springer Nature Switzerland AG 2022
N. Urquhart, *Nature Inspired Optimisation for Delivery Problems*,
Natural Computing Series, https://doi.org/10.1007/978-3-030-98108-2_8

- *dist(Node n1, Node n2)*
  returns the distance between n1 and n2, if they are linked by a Way.
- neighbours(Node n)
  returns a list all the nodes that are directly linked to n.
- *allNodes( )*
  returns a list of all the nodes in the graph.
- *getWay(Node n1, Node n2)*
  returns a Way object if n1 and n2 are linked directly, else returns null.

Dijkstra

The first implementation of a route finding algorithm was published in 1959 (although it was developed in 1956) by Edsger W. Dijkstra. Several variants of the algorithm exist, one that carries out an exhaustive search, finding the distance and route from a start node to every other node in the graph—in effect, finding the shortest path tree. Reputedly, this variant was used to find the distance from Amsterdam Central Station to every other railway station in the Netherlands. A pseudocode implementation of this algorithm may be seen in Algorithm 16 and a worked example of the algorithm may be seen in Fig. 8.1.

Each node in the graph has two labels, one to record the distance from the start node and one to hold the identity of the previous node (the node from which the distance was calculated). Initially, all the distances are set to infinity and the previous labels are set to null (lines 2 and 3). Dijkstra's algorithm maintains a list of unvisited nodes (line 6). The algorithm now executes a loop, considering each unvisited node (lines 7–15). Within the loop, the neighbours of the current node are found (line 8), and the distance to each neighbour is calculated as the distance to the current node plus the extra distance between the current and the neighbour (line 10). If the newly calculated distance is less than that on the neighbours' label (line 11), then we update the distance and previous labels. This ensures that if there is more than one route from the start node to any other node; the route that results in the shortest distance is recorded. After each neighbour has been considered, all of the neighbours are removed from the unvisited list (line 17). Finally, the unvisited nodes are sorted in distance label order and the one with the least distance is removed from the unvisited list and becomes the new current node (lines 14–15). Figure 8.1a shows the graph in its initial state, with the start node A designated as current and all other nodes unvisited. Nodes B and D are the direct neighbours of A so they are considered next; their labels are updated and node B (least distance from A) becomes the new current node (Fig. 8.1b). At the next stage (Fig. 8.1e), nodes C and D are considered (as neighbours of B), and the labels of C are updated, but as D already has a distance (A–D) that is shorter than A–B–D the labels on D are left unchanged. The algorithm continues to iterate until the unvisited list is empty (Fig. 8.1f). At this point, every node is labelled with the distance from the start (A). To find the route from the

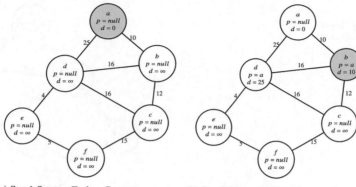

(a) Step1:Start=c End=c Current=a
Unvisited=[bcdef]

(b) Step2:Start=a End=c Current=b
Unvisited=[cdef]

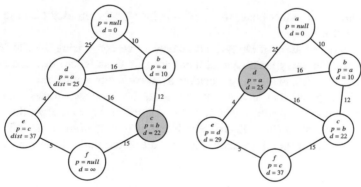

(c) Step3:Start=a End=c Current=c
Unvisited=[def]

(d) Step4:Start=a End=c Current=d
Unvisited=[ef]

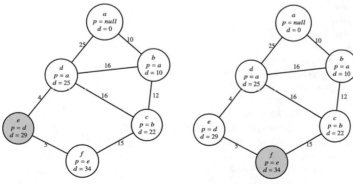

(e) Step5:Start=a End=c Current=e
Unvisited=[f]

(f) Step6:Start=a End=c Current=f
Unvisited=[]

Fig. 8.1   Dijkstra's algorithm

Fig. 8.2 The route A–C
established by using the Prev
label for each node

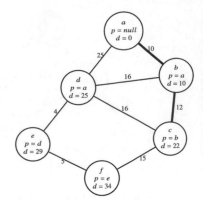

finish node (or any other node) to A, the previous labels are used (see Fig. 8.2 and Algorithm 18).

This implementation of Dijkstra constructs a tree comprising the shortest route from the start to every node. Whilst there are situations where this is useful (e.g. calculating the distance from one station to every other station on a network), it is overkill when all that is required is a route between two nodes. A modified implementation of Dijkstra may be found in Algorithm 17. This algorithm contains a simple modification (lines 15–16) which causes the algorithm to terminate when the current node is the finish node, thus not all of the nodes have their distances calculated.

Algorithm 16 Dijkstra's algorithm—shortest route tree variant.

1: **Procedure** $Dijkstra(start)$
2: $dists[] = infinity$
3: $previous[] = null$
4: $current = start$
5: $dists[current] = 0$
6: $unvisited = Graph.allNodes()$
7: **while** $unvisited.size > 0$ **do**
8:     $neighbours = Graph.neighbours(current)$
9:     **for** $Node in neighbours$ **do**
10:         $distToN = dists[current] + Graph.dist(current, n)$
11:         **if** (**then** $distToN < dists[n]$)
12:             $dists[n] = disToN$
13:             $previous[n] = current$
14:     $unvisited.remove(current)$
15:     $current = findSmallest(dists, unvisited)$

Algorithm 17 Dijkstra's algorithm (modified).

1: **Procedure** $DijkstraMod(start, finish)$
2: $dists[] = infinity$
3: $previous[] = null$
4: $current = start$
5: $dists[current] = 0$
6: $unvisited = Graph.allNodes()$
7: **while do**$(unvisited.size > 0)$
8: $\quad neighbours = Graph.neighbours(current)$
9: $\quad$ **for do**$(Node in neighbours)$
10: $\quad\quad distToN = dists[current] + Graph.dist(current, n)$
11: $\quad\quad$ **if then**$(distToN < dists[n])$
12: $\quad\quad\quad dists[n] = disToN$
13: $\quad\quad\quad previous[n] = current$
14: $\quad\quad$ **if then**$(current == finish)$
15: $\quad\quad\quad return$
16: $\quad unvisited.remove(current)$
17: $\quad current = findSmallest(dists, unvisited)$

Algorithm 18 Print route details.

1: **Procedure** $PrintRoute(start, finish, prev)$
2: $dist = 0$
3: $current = finish$
4: $old = previous[finish]$
5: **while do**$(current! = null)$
6: $\quad myWay = Graph.getWay(current, old)$
7: $\quad print(myWay.name())$
8: $\quad dist = dist + Graph.getDist(old, current)$
9: $\quad old = current$
10: $\quad current = previous[current]$
11: $print("Distance = " + dist)$
12: **EndProcedure**

A*

Dijkstra's algorithm potentially has a long run time, which increases with the number of nodes in the graph. The need to search the unvisited list at each step can adversely affect performance. This property can make Dijkstra unattractive for use with large graphs.

An alternative to Dijkstra is the A* (pronounced "A-Star") algorithm Hart et al. (1968). A* uses a data structure called the *Open List* to hold only those nodes that are being considered and directs the search using a heuristic to determine which node should become the current node. When treating the graph as a representation of a road network, the heuristic can be based upon the direct distance from the current node to the finish node. In our case, the heuristic used is the great circle distance (calculated using the Haversine formula (see Sect. 7.3) to the finish node. This great circle distance is only an estimate, as road networks rarely follow the exact straight line between locations.

Algorithm 19 shows the basic A* code; the nodes are labelled with distance and previous in the same manner as Dijkstra. The Open list is initialised with the start node (line 6). Within the main loop (lines 7–19), the neighbours of the current node are obtained (line 8). Each neighbour is then considered in turn (line 9), and the distance from the start to that node is calculated in the same manner as Dijkstra (line 10). If the distance to the neighbour is an improvement on any previous distance (the nodes are all initialised with a default distance label of infinity), then the node is added to the open list - providing that it is not already in the list (lines 12–13). After the neighbours have each had their distances re-calculated and their labels updated (where appropriate), the current node is removed from the open list and a new current node is determined. The new current node is determined by applying the heuristic to each node in the open list and returning the node with the lowest value (Algorithm 20). In this case, the heuristic used is

$$h = ds + de$$

where
ds is the known distance from the start node.
de is the estimated distance to the finish node (calculated using the Haversine formula).

In this way, the selected node will be the one that appears to lie on the most direct route between the start and the finish nodes. Figure 8.3 illustrates a simple example of A* in use.

Algorithm 19 A-Star.

```
1: Procedure AStar(start, finish)
2: dists[]= infinity
3: previous[]= null
4: current = start
5: dists[current]= 0
6: open.add(current)
7: while do(open.size > 0)
8:    neighbours = Graph.neighbours(current)
9:    for do(Nodeninneighbours)
10:       distToN = dists[current]+Graph.dist(current, n)
11:       if then(distToN < dists[n])
12:          if then(!open.contains(n))
13:             open.add(n)
14:          dists[n]= disToN
15:          previous[n]= current
16:    open.remove(current)
17:    current = findClosest(dists, open)
18:    if then(current == finish)
19:       return
```

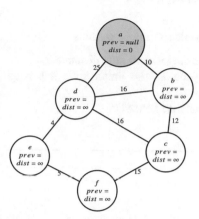

(a) Step 1: The initial state, a is added to the open list. openlist=$a = 0$

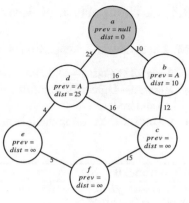

(b) Step2: The neighbours of a are added to the open list. The heuristic $(ds + de)$ is used to order the list. $openlist = d = 28(25 + 3), b = 29(10 + 19)$

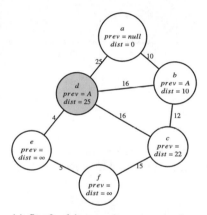

(c) Step3: d is now the current node being removed from $openlist$. c and s are added to $openlist$. $openlist = [b = 29(10+19), c = 32(22+10), e = 29(29 + 0)]$

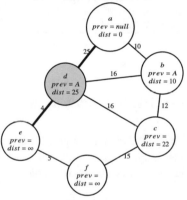

(d) Step4: As the finish node (E) has been reached and re-labeled, the algorithm can now halt and use the prev labels to trace the desired route from A to E.

Fig. 8.3 A Simple example of A* in use

8.2 A Comparison of Basic Search Algorithms

We will undertake a comparison of a group of basic routing algorithms:

- **Floodfill Dijkstra**—Dijkstra's algorithm returning a minimum distance tree.
- **Dijkstra**—Dijkstra's algorithm halting as soon as the finish node has been encountered.
- **A-Star**—Search using the minimum direct distance heuristic.

Algorithm 20 A-Star—heuristic-based selection

1: **Procedure** $findClosest(dists, unvisited)$
2: $closestD = infinity$
3: $closestN = null$
4: **for do**$(Node n in unvisited)$
5:    $estim = dists[n] + EuclideanDist(n, finish)$
6:    **if then**$(estim < closestD)$
7:       $closestD = estim$
8:       $closestN = n$
9: Return $closestD$

Java Implementation

A basic Java routing engine for demonstrating routing algorithms is shown in Fig. 8.4. As with previous examples in this book, the Java code is available for download from the repository that accompanies this book.

Parsing OpenStreetMap Data

In order to test the algorithms, we utilise data from the OpenStreetMap database OpenStreetMap contributors (2021) (see listing 8.1). OSM utilises a comprehensive XML-based vector format (see Fig. 7.4) which contains far more data than we need for simple routing. We use the *BasicOSMParser* library Pavie (2021) to read the .OSM files (line 12). Our implementation makes use of the OSM parser *info.pavie.basicosmparser* which provides the means of extracting elements (e.g. Ways and Nodes) from the .OSM file, which are stored in a *Map* object (line 12).

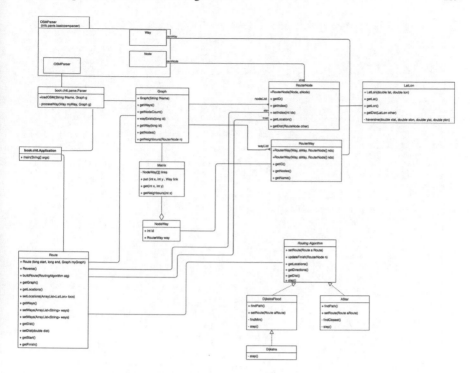

Fig. 8.4 The demonstration routing engine described in this chapter

Within the OSM scheme, a Way element describes a *polyline* that connects a series of Node elements. A Way element can represent many entities including roads, railways, coastlines, bus routes, etc. In this case, we are only interested in those that define highways; they can be distinguished within the OSM data as their IDs are prefixed with a "W" (line 19). Each way element is then processed (lines 31–56) and a *RouterWay* object is created. Ways which do not have a *Highway* attribute (line 34) are filtered out. We define a set of Highway types that we are not interested in (line 3), and filter out any ways belonging to these highway types (line 37). By removing highway types that we do not wish to include in our routes, we can reduce the size of our graph.

```
1
2    private static ArrayList<String> inValidHighways = new ArrayList<
         String>(
3      Arrays.asList("cycleway","elevator","bus_stop","proposed","no",
4      "path","corridor","steps","bridleway","footway","raceway",
5      "escape","bus_guideway","pedestrian","construction"));
6
7    public static void loadOSM(String fName, Graph graph) {
8      //Load the contents of <fName> into <graph>
9      OSMParser p = new OSMParser();      //Initialise parser
10     try {
11       System.out.println("Loading Ways");
12       Map<String,Element> result = p.parse(new File(fName));
13       //Parse OSM data, and put result in a Map object
14
15       for (String s : result.keySet()){
16         Element e = result.get(s);
17         //Filter nodes
18         String id = e.getId().trim();
19         if (id.startsWith("W")){
20           Way w = (Way)e;
21           RouterWay newR = processWay(w,graph);
22           if (newR != null)
23             graph.addWay(newR);
24         }
25       }
26     } catch (IOException | SAXException e) {
27       e.printStackTrace();//Report any parse errors
28     }
29   }
30
31   private static RouterWay processWay(Way myWay, Graph g){
32     //Create a RouterWay object based upon the <way> object
33     String  highway = (String)myWay.getTags().get("highway");
34     if(highway==null)
35       //Any ways that do not have highway attribute should be
         disregrded.
36       return null;
37     if (inValidHighways.contains(highway ))
38       //Disregard any highways types that are in our 'invalid list'
39       return null;
40
41     ArrayList<RouterNode> nodes = new ArrayList<RouterNode>();
42     //add nodes associated with myWay to the RouterWay object
43
44     for (Node n : myWay.getNodes()){
45       long nid = convertId(n.getId());
46       if (!g.nodeExists(nid)){//Check node has been loaded.
47         //Add node
48         g.addNode(new RouterNode(n));
49       }
50       RouterNode rn = g.getNode(nid);
51       nodes.add(rn);
52     }
53     RouterNode[] myNodes =  nodes.toArray(new RouterNode[nodes.size
         ()]);
54     RouterWay newWay = new RouterWay(myWay, myNodes);
55     return newWay;
56   }
```

Listing 8.1 Parsing the OSM file

Creating a street graph in memory

We define a *Graph* class (see the class diagram in Fig. 8.4) which contains our data structure. We use the *RouterNode* and *RouterWay* objects to hold details of OSM nodes and ways. In order to speed up graph operations, we use a *Matrix* class to hold details of which nodes are directly linked by a way.

Within the Matrix class (see listing 8.2), the obvious internal data structure is a 2-dimensional array of size n by n where node x and node y are linked; then the matrix entry at $[x, y]$ is a reference to the *way* object that links x and y. This is based on the scheme outlined earlier in Sect. 7.4.

If we consider the case of the City of Edinburgh, Scotland, using the OSM data for Edinburgh would result in a graph of 63267 nodes. Using a matrix to store direct links would require an array of 63267^2 (or 4002713289) elements. If we assume that a reference in Java takes 32 bytes (the memory required may vary depending on the JVM version used) that would equate to over 16 gigabytes of memory, a high percentage of these array elements will contain a null value as most nodes are not directly linked.

Our implementation uses a the *NodeWay* class that stores a node index and a pointer to a *RouterWay* object. Each node in the graph has rows in the *links* array (line 7). Each row is initialised to a size of 10 (line 16) allowing for up to 10 ways to be associated with that node (this can be increased by re-sizing the row if needed—see below). This equates to a default size of 2.5 Mb for the Edinburgh graph, a considerable saving on the n by n implementation. A new link can be added to the matrix using the *put* method (lines 20–35) which includes a provision for expanding the row if the number of neighbours exceeds the default 10 (lines 33–34). The *get* method (lines 39–48) takes two node indexes and returns the appropriate *RouterWay* object if the two nodes are directly linked, else it returns *null*. The most useful method within Matrix is the *getNeighbours* method (lines 50–60), which accepts a node index and returns an array comprising all of the *RouterWay* objects associated with that node.

```
 1  public class Matrix {
 2    private class NodeWay{
 3      public int id;
 4      public RouterWay way;
 5    }
 6
 7    private NodeWay[][] links;
 8
 9    public Matrix(int size){
10      //Initialise a matrix for a graph with <size> nodes;
11      //Note that NodeWay is not sizexsize, but is initially
12      //sizex10. The node IDs used here are the position of
13      //the node within the Graph object and NOT the OSM ids.
14      links = new NodeWay[size][];
15      for (int c= 0; c < size; c++){
16        links[c] = new NodeWay[10];
17      }
18    }
19
20    public void put(int x, int y, RouterWay w){
21      //Aad a link between node<x> and node<y> via Way <w>
22      NodeWay nw = new NodeWay();
23      nw.id = y;
24      nw.way = w;
25      //Find the first null entry in the matrix at row <x>
26      int oldSize = links[x].length;
27      for (int i=0; i < oldSize;i++)
28        if (links[x][i]==null){
29          links[x][i] = nw;
30          return;
31        }
32      //if there's no space in row <x> then resize the row by 5.
33      links[x] = Arrays.copyOf(links[x], oldSize + 5);
34      links[x][oldSize+1] = nw;
35    }
36
37    //If nodes <x> and <y> are linked by a way, return a reference
38    //to that way, if the nodes are not directly linked, return null.
39    public RouterWay get(int x, int y){
40      NodeWay[] row = links[x];
41      for(NodeWay nw : row){
42        if (nw!=null){
43          if (nw.id==y)
44            return nw.way;
45        }
46      }
47      return null;
48    }
```

Listing 8.2 The Matrix class implementation

Algorithm implementation

As with the earlier examples of TSP and VRP solvers, utilising object-oriented techniques to ensure that the implementation of routing algorithms is not visible to objects making use of the algorithms. We implement an abstract class (*RoutingAlgorithm*) and have our algorithms' sub-class RoutingAlgorithm; this allows the implementation of the algorithm (e.g. Dijkstra or A*) to be independent of the main application.

All of the algorithms used are iterative; *RoutingAlgorithm* has a method *step()* which undertakes one iteration of the algorithm. The use of the *step()* method allows the BiDirectional algorithms (*DijkstraBiDirectional* and *AStarBiDiDirectional*) to be implemented using the existing *Dijkstra* and *AStar* classes as in listing 8.4. Two *RoutingAlgorithm* objects are declared (lines 1 and 2); the *findPath()* method (lines 5–24) calls the *step()* method of the *forward* and *reverse* algorithms and compares (lines 14 and 15) the nodes currently being considered by the two algorithms; when they have a node in common, the search halts.

```
1    public abstract void findPath();//Find a path from start to
2            //finish, by calling step() repeatedly
3    public abstract Object step();//Run one step of the algorithm
4
5    public void setRoute(Route aRoute){
6        //aRoute contains details of the path to be found.
7        this.theRoute = aRoute;
8        this.myGraph = aRoute.getGraph();
9        this.start = aRoute.getStart();
10       this.finish = aRoute.getFinish();
11   }
12
13   public void setFinish(RouterNode f){
14       finish = f;
15   }
16
17   public ArrayList<LatLon> getLocations(){
18       //Return a list of LatLon objects that represent the path
19       ArrayList<LatLon> res = new ArrayList<LatLon>();
20       RouterNode current = finish;
21       while (current != start){
22           res.add(res.size(),current.getLocation());
23           //add at end to reverse order
24           current = previous[current.getIndex()];
25       }
26       return res;
27   }
28
29   public ArrayList<String> getRoadNames(){
30       //Return the  road names that comprise the path found
31       ArrayList<String> res = new ArrayList<String>();
32       RouterNode old = null;
33       RouterNode current = finish;
34       while (current != start){
35           RouterWay currentWay = myGraph.getWay(current,old);
36           if (currentWay != null)
37               res.add(res.size(),currentWay.getName());
38           //Add in reverse order
39           old = current;
40           current = previous[current.getIndex()];
41       }
42       return res;
43   }
```

Listing 8.3 An extract from the abstract class RoutingAlgorithm

```
1   private Dijkstra   forward;
2   private Dijkstra   reverse;
3
4   @Override
5   public void findPath() {
6     boolean done = false;
7     forward.current = this.start;
8     reverse.current = this.finish;
9     while(!done){
10      ArrayList<RouterNode> fCurrent = forward.step();
11      ArrayList<RouterNode> rCurrent = reverse.step();
12      //Look for a common node in the nodes currently
13      //being considered by forward and reverse
14      for (RouterNode join: fCurrent) {
15        if (rCurrent.contains(join)){
16          //A complete route has been found
17          forward.setFinish(join);
18          reverse.setFinish(join);
19          done = true;
20          break;
21        }
22      }
23    }
24  }
```

Listing 8.4   An extract from the *DijkstraBiDirectional* class showing how two *Dijkstra* objects are used for the forward and reverse searches

Creating a Routing Application

We use the *Route* class (see listing 8.6) as a Facade to our routing engine. The usage of *Route* is demonstrated in listing 8.5. Note that we can easily change the routing algorithm used (line 6).

This simple routing engine has a number of limitations:

- No error handling exists to cope with situations where the graph is incomplete.
- Routes are only found from Node to Node; the user needs to specify the OSM node ids.

```
1   Graph myGraph = new Graph("westscot.osm");
2   //Load osm data into a street graph
3   Route testRoute = new Route(myGraph,291781127L,257927392L);
4   //Create a Route object within the graph based on the start and
      end nodes
5   testRoute.buildRoute(new AStar());
6   //Find a path between the start and the end using A*
7   System.out.print("distance," + testRoute.getDist());
```

Listing 8.5   An example of a route distance being calculated using the code developed for this chapter

```
1  public class Route {
2    private RouterNode start;
3    private RouterNode finish;
4    private ArrayList<LatLon> locations = new ArrayList<LatLon>();
5    private ArrayList<String> ways = new ArrayList<String>();
6    private double dist=0;
7    private Graph myGraph;
8
9    public Route(Graph myGraph, long start, long finish){
10     //Find a path from <start> to <finish> through <myGraph>
11     if (!myGraph.nodeExists(start)){
12       System.out.println("Start not found");
13       System.exit(-1);
14     }
15
16     if (!myGraph.nodeExists(finish)){
17       System.out.println("Finish not found");
18       System.exit(-1);
19     }
20
21     this.myGraph = myGraph;
22     this.start = myGraph.getNode(start);
23     this.finish = myGraph.getNode(finish);
24   }
25
26   public Route reverse(){
27     //Create a new Route object that is the Reverse of this one
28     return new Route(this.myGraph, this.getFinish().getId(), this.
         getStart().getId());
29   }
30
31   public void buildRoute(RoutingAlgorithm algorithm){
32     //Find the desired path using <algorithm>
33     algorithm.setRoute(this);
34     algorithm.findPath();
35     dist = algorithm.getDist();
36     locations = algorithm.getLocations();
37     ways = algorithm.getRoadNames();
38   }
```

Listing 8.6   An extract from the Route class

Results

We test our routing algorithms on two networks:

- **Sirmione, Italy** The commune of Sirmione is located on the Sirmio peninsula which projects into Lake Garda in northern Italy. The street graph has 255 nodes, connected by 60 ways.
- **Edinburgh, UK** The City of Edinburgh is the capital of Scotland, with a population of 500,000. The street graph contains 63267 nodes, connected by 10612 ways.

Table 8.1 gives the results obtained on the small Sirmione graph. We note that there is very little difference in performance between the algorithms in terms of time and distance. When the graph is relatively small, the time advantages of using A*

Table 8.1 The results obtained on the Sirmione set of short journeys. Distances are in Km and time ms

Problem ID	DijkstraFlood		Dijkstra		A-Star	
	Distance	Time	Distance	Time	Distance	Time
1	0.94	2	0.96	2	0.94	5
1R	0.94	1	0.94	2	0.94	1
2	0.80	1	0.80	1	0.80	1
2R	0.80	1	0.80	1	0.80	0
3	0.62	1	0.62	0	0.62	1
3R	0.62	1	0.62	0	0.62	0
4	0.94	1	0.94	1	0.94	1
4R	0.94	1	0.96	0	0.94	0
5	0.80	1	0.80	1	0.80	0
5R	0.80	0	0.80	0	0.80	0
6	0.10	1	0.10	0	0.10	0
6R	0.10	1	0.10	0	0.10	0

are minimal. In a scenario such as this, we note that DijkstraFlood takes between 1 and 2 milliseconds to execute. As DijkstraFlood calculates the distance from the start node to every other node, executing it once for each node would allow every possible journey within the graph to be calculated:

$$t = n * e$$

where
t is the total time required.
n is the number of nodes.
e is the execution time (ms) for DijkstraFlood.

For the Sirmione scenario, the total time to calculate every possible distance is 510ms; it would therefore make sense in a simple example such as this to calculate an OD matrix once and then look up the matrix when solving a problem.

Table 8.2 shows the results obtained over 28 routes within the Edinburgh City dataset (see Sect. 8.2). Each route was constructed 10 times in order to establish an average time taken; the distances obtained did not change over each run.

Figure 8.5 plots the distances obtained on the 28 Edinburgh routes (each route is searched in both directions). The results appear similarly, so we apply a T-test to determine if there is any statistically significant difference in results. Where a T-test produces a value of p greater than 0.9, it can be accepted that there is no statistical significance. In this example in an urban area, no difference in results is noted between DijkstraFlood (optimal) and Dijkstra ($p = 9.571$) and DijkstraFlood and A*($p = 1.0$).

Table 8.2 Routing algorithm performance across 28 routes within the Edinburgh graph. Times are an average of 10 runs

Problem ID	DijkstraFlood		Dijkstra		Astar	
	Distance	Time	Distance	Time	Distance	Time
1	4.57	6823	4.57	4763	4.57	94
1R	4.56	6384	4.56	5617	4.56	166
2	6.64	6869	6.64	6203	6.64	542
2R	6.64	7814	6.64	6152	6.64	694
3	2.07	7279	2.07	1259	2.07	14
3R	2.07	7397	2.08	1345	2.07	32
4	3.13	7113	3.13	2424	3.13	27
4R	3.13	7387	3.14	3000	3.13	125
5	1.29	7536	1.29	554	1.29	10
5R	1.29	8088	1.29	970	1.29	12
6	4.89	6713	4.89	5927	4.89	370
6R	4.68	7058	4.68	5015	4.68	211
7	1.69	6309	1.69	925	1.69	15
7R	1.69	6766	1.69	1146	1.69	43
8	1.94	7726	1.94	997	1.94	7
8R	2.02	8123	2.02	1416	2.02	22
9	3.47	7215	3.72	2698	3.47	122
9R	3.47	6534	3.47	3585	3.47	72
10	5.73	6813	5.73	6164	5.73	436
10R	5.73	6988	5.73	5860	5.73	228
11	6.16	6640	6.18	6199	6.16	500
11R	6.16	7549	6.16	5974	6.16	120
12	4.54	6564	4.54	6098	4.54	249
12R	4.54	9283	4.54	6421	4.54	153
13	4.86	6792	4.86	6391	4.86	461
13R	4.86	8217	4.86	6272	4.86	551
14	3.88	7522	3.88	4846	3.88	85
14R	3.88	6517	3.88	4067	3.88	42
15	2.47	6984	2.47	1737	2.47	45
15R	2.47	6533	2.48	1345	2.47	48
16	3.15	7496	3.33	2670	3.15	181
16R	3.15	6208	3.32	3226	3.15	117
17	7.60	7360	7.60	8795	7.60	1258
17R	7.60	7315	7.60	7068	7.60	808
18	7.71	6620	8.04	6514	7.71	1289
18R	7.71	8599	7.71	7218	7.71	1739

(continued)

Table 8.2 (continued)

Problem ID	DijkstraFlood		Dijkstra		Astar	
	Distance	Time	Distance	Time	Distance	Time
19	6.41	6697	6.41	6934	6.41	754
19R	6.41	8407	6.41	9367	6.41	407
20	3.97	7216	3.97	3543	3.97	79
20R	3.97	7836	3.97	3393	3.97	85
21	5.75	7664	5.75	6773	5.75	682
21R	5.75	6510	5.75	5436	5.75	250
22	1.85	9018	1.85	2722	1.85	8
22R	1.85	6505	1.85	1408	1.85	28
23	4.14	6927	4.14	4162	4.14	74
23R	4.14	7210	4.30	4143	4.14	90
24	1.22	7805	1.22	164	1.22	2
24R	1.22	7998	1.22	628	1.22	7
25	1.23	7611	1.23	765	1.23	4
25R	1.23	8583	1.23	1690	1.23	3
26	5.23	8459	5.23	5330	5.23	485
26R	5.23	6999	5.23	5926	5.23	427
27	4.11	7369	4.11	5302	4.11	138
27R	4.06	6772	4.06	4947	4.06	159
28	0.66	7298	0.66	266	0.66	1
28R	0.66	6507	0.66	241	0.66	4

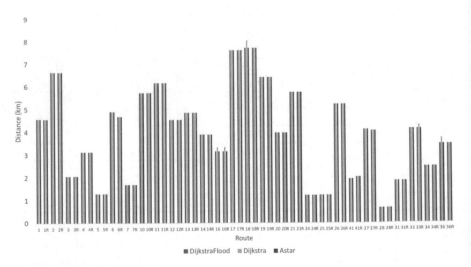

Fig. 8.5 The distances of the routes found within the Edinburgh dataset

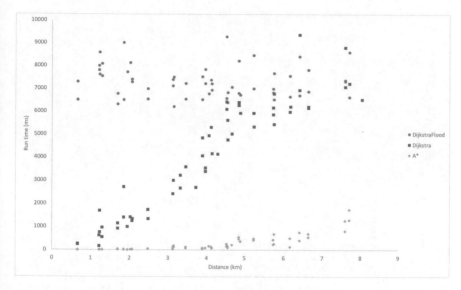

Fig. 8.6 Distance versus execution time for the DijkstraFlood, Dijkstra and A* algorithms

Figure 8.6 plots the average execution time for each of the 28 test routes against the length of the route found. The DijkstraFlood algorithm takes the longest to execute, but finds the optimum. The time taken to execute DijkstraFlood should be constant as it is governed by the size of the graph as each node is included in the search. The times shown are not equal for each run; this is due to the nature of the Java virtual machine (which can execute the garbage collection process as required), and the operating system (MS Windows 10) will alter the CPU availability to an application based on other applications and associated interrupts. Note that DijkstraFlood always takes 6000–9000 ms; as discussed, the variance in timings is due to the interaction with background processes and is not related to the distance. The modified Dijkstra algorithm shows a definite correlation between execution time and distance, essentially the shorter the route the less of the graph is searched. When we consider A*, we note that the run times are significantly less than either of the Dijkstra variants. Although the run times are less, A* does show a link between route distance and run time.

Of the algorithms examined (Dijkstra, DijkstraFlood and A*), it becomes apparent that A*, in this case, finds routes that are the same length as that found by Dijkstra. If we examine the route found (see Fig. 8.7), we note that the route taken is close to the straight line route between the start and end. If the road network allows a route that is close to the straight line, then that will favour the Euclidean distance heuristic.

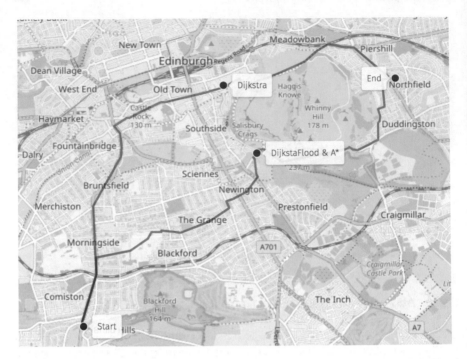

Fig. 8.7 An example (route 18 in Table 8.2) showing the identical routes found by DijkstraFlood and A*. The slightly longer route found by Dijkstra is also shown. When the route is searched for in the opposite direction, all three algorithms find the same route. *Base map and data from OpenStreetMap and OpenStreetMap Foundation* https://www.openstreetmap.org/copyright

The West of Scotland Example

A* has a significant drawback in situations where the desired route is not largely aligned with the direct route from start to finish. Consider the problem shown in Fig. 8.8, to find a route from Glasgow to Campbeltown. Because of the geography of the West of Scotland, it is necessary to head north-west from Glasgow when travelling to Campbeltown, despite Campbeltown being located south-west of Glasgow. The reason is the Firth of Clyde and sea lochs which have to be circumnavigated. In this scenario, the A* search will follow a route south-west, in the direction of the dotted line. Although the dotted line runs directly to Campbeltown, it becomes apparent that a large body of water (the Firth of Clyde to be exact) prevents progress by road in this direction. The actual route via the road network requires the traveller to head north for a considerable distance in order to reach Campbeltown. In such situations, the advantage of A* may be lost.

Table 8.3 shows the results obtained on the Glasgow to Campbeltown problem. Only algorithms which could return a result within 20 min were considered; the times presented are an average of over 10 runs. The only algorithms which could return

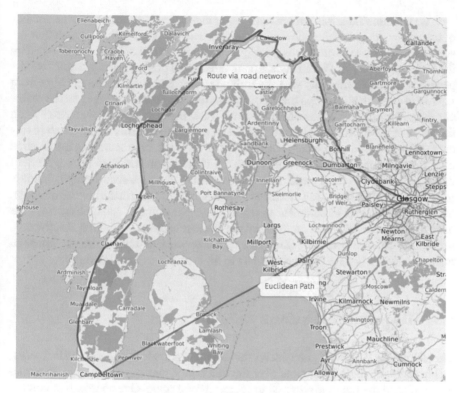

Fig. 8.8 The Glasgow to Campbeltown problem. The red line shows the Euclidean route that the A* search will attempt to follow; the blue line shows the actual route. *Base map and data from OpenStreetMap and OpenStreetMap Foundation* (https://www.openstreetmap.org/copyright)

Table 8.3 The results obtained from the West of Scotland dataset (Glasgow to Campbeltown). Note that in both cases, the routing takes significantly longer when searching from the Glasgow end of the route

	Dijkstra			A*		
	Dist	Time	Nodes Searched	Dist	Time	Nodes Searched
Glasgow to Campbeltown	196.82	2547435	779346 (93%)	196.82	138,905	709774 (84%)
Campbeltown to Glasgow	196.94	5162241	262468 (31%)	196.94	414144	107944 (13%)

a result within 20 min were Dijkstra and A*. Both algorithms find identical routes, albeit with slight differences in each direction. The graph has a total of 842315 nodes and 117447 ways; the graph is at its most dense around the City of Glasgow (middle left of the map), whilst the areas on the right represent rural areas and have a very sparse road network.

If we examine Table 8.3, we can see a major indication of performance, the number of nodes considered during the execution of the algorithm. Firstly, we note that A* searches fewer nodes, which is as we might expect given that the search is directed by the Euclidean heuristic. When considering directional performance, we note that A* performs far more efficiently when searching to Glasgow then when searching from it. We can explain this through the graph structure, when commencing from the centre of Glasgow the A* search is directed south-west; this becomes a "dead end" when the water of the Firth of Clyde is reached. Because of the density of the graph in this area, A* will continue to search possibilities in this area, only searching northwards after covering many other options. Once the route North has been found, A* will find the southbound route easier as it is in a rural area with a sparse graph, and it is following mostly a Euclidean distance to the finish. Searching in the reverse direction from Campbeltown, A* commences with a very sparse graph, and quickly finds the only main road northwards on the Kintyre Peninsula; when searching southwards, the A* algorithm is searching through a dense network and hence the performance is impressive; only 13% of the graph has to be searched to find the route.

8.3   Hierarchical Structures

The size of the OSM street graph increases as the geographical area covered increases; this leads to practical constraints over the amount of memory required to construct the graph and the run time required to search it for a route. One approach to combat these issues is to adopt a hierarchical approach to route finding.

Road networks generally follow a hierarchy, at the lowest end are residential streets, which progress through levels of importance until, at the top of the hierarchy are motorways (or autobahns, freeways, etc.). The hierarchy of highways in OpenStreetMap is as follows:

Way type
Motorway
Primary
Secondary
Tertiary
Unclassified
Residential

Most journeys commence on roads at the lower end of the hierarchy and progress along roads that of types that are higher within the hierarchy, before descending through the hierarchy as the journey completes. Figure 8.9 shows the classification of each highway traversed on a journey from the City of Edinburgh to the City of Glasgow. Note how the journey begins and ends on roads that are lower within the hierarchy, and spends the middle section at the top of the hierarchy. As most journeys follow this pattern, we can exploit this when building a routing application.

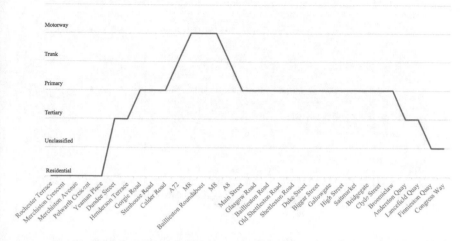

Fig. 8.9 A journey from the centre of Edinburgh to the centre of Glasgow, showing the classification of each highway. Note this chart does not indicate the distance travelled; in reality, the motorway section comprises most of the journey

Dividing into Hierarchies

In our examples so far in this chapter, we have had all of our roads contained within one street graph, but we can split that graph by highway level into sub-graphs. For instance, we could split our street graph into 3 levels; see Table 8.4. We now have 3 graphs each containing part of the street graph. Only the level 0 graph will be fully connected, levels 1 and 2 will comprise a series of ways and nodes that are not fully connected. Figure 8.10 shows how our journey progresses through layers in each graph. There is also a vertical aspect to these graphs, as each pair (0, 1 and 1, 2) will have common nodes, e.g. where a primary road intersects with a secondary road both graph 0 and graph 1 will share a common node.

Algorithm 21 shows a very simple 3-tier algorithm. The array *level[]* contains 3 graphs (as per Table 8.4). The function *grapth.findLink(node)* returns the nearest Node to <node> that is a link up to the next layer. The *findLink()* function can use a

Table 8.4 Grouping Ways by highway type

Graph	Way type
0	Motorway
	Primary
1	Secondary
	Tertiary
2	Unclassified
	Residential

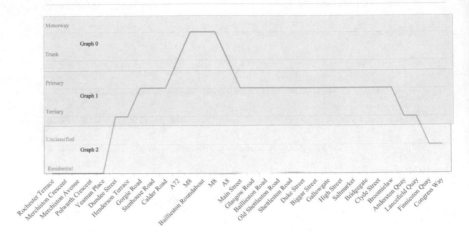

Fig. 8.10 The journey illustrated in Fig. 8.9 with the 3-tier graph structure superimposed

modified version of Dijkstra which terminates when a link node has been found and *path( )* may make use of any of the routing algorithms described earlier.

There are three principal advantages to this approach:

- **Parallelism** The searches from the start (lines 2 and 3) and from the end (lines 4 and 5) may be conducted in parallel, possibly on different machines.
- **Caching** It is likely that a number of routes used on the lower levels (e.g. level 0) will be used frequently (e.g. links between major cities); such frequently used links may be cached to avoid them being re-calculated every time.
- **Graph Size** By dividing the graph by level at no point does any routing algorithm have to search through the entire graph.

Algorithm 21 Outline hierarchy algorithm.

1: **Procedure** hierarchy(start, end)
2: routeSx = level[2].findLink(start)
3: routeSy = level[1].findLink(routeSx.end)
4: routeEx = level[2].findLink(end)
5: routeEy = level[1].findLink(routeEx.end)
6: routeMiddle = level[0].path(routeSy.end,routeEy.end)
7: result = routeSx + routeSy + routeMiddle + routeEy + routeEx
8: **Return** result

A Simple Java Implementation

We can implement a simple 2-tier hierarchy in Java making use of the framework and algorithms developed earlier in this chapter. Listing 8.8 shows the *main( )* function, the graphs are created (line 7 and listing 8.7) and the system is tested by generating routes between random points.

The creation of the graphs is shown in listing 8.7; the *level[0]* graph only contains motorway, motorway_link, trunk and trunk_link ways; everything else is excluded (line 5). Note that level 0 includes motorway and trunk routes as the motorway network on its own does not create a fully connected graph. Prior to loading the data, the *OSMFilter* utility Weber (2020) was used to remove all data other than the road network from the .OSM file in order to reduce the memory requirements of the parser.

```
1    Graph res= new Graph(fName,new ArrayList<String>( Arrays.
2      asList("motorway","motorway_link","trunk","trunk_link")));
3    System.out.println(fName + " loaded "  +res.getNodeCount());
4    return res;
5    }
6  }
```

Listing 8.7 Setting up the levels

```
1  public class HierarchyTest {
2    private static final int MAX_LEVELS = 2;
3    private static Random rnd = new Random();
4    static private Graph[] levels = new Graph[MAX_LEVELS];
5
6    public static void main(String[] args) {
7      setup("./data/roads.osm");//Load the graphs
8
9      int count =0;
10     while (count <100){
11       //Test on 100 random routes
12       System.out.println("Test "+count);
13       long start = getRandom(1).getId();
14       long end = getRandom(1).getId();
15       Route r = buildRoute(start, end);
16       if (r != null){
17         /*r is null if the route cannot be
18          * built for some reason.
19          */
20         KMLWriter out = new KMLWriter();
21         out.addRoute(r.getLocations(), "", "", "red");
22         out.writeFile(count+".kml");
23       }
24       count++;
25     }
26   }
27
28     private static RouterNode getRandom(int level){
29       /* Return a random node from the graph a
30        * <level>.
31        */
32       return (RouterNode) new ArrayList(levels[level].getNodes
         ()).get(rnd.nextInt(levels[level].getNodes().size()));
33     }
```

Listing 8.8 Demonstrating the use of a hierarchical graph structure

The *buildRoute* method (listing 8.9) shows the construction of the route. Lines 9–11 show the construction of the first leg through level [1]. The router used is *DijkstraFindUpLink* which is a variant of Dijkstra which halts as soon as it encounters a node that is linked to the next layer. The same process is undertaken for the final leg (lines 12–15).

Having identified the two *Uplink* nodes, a check is made to ascertain if they are the same node (line 17). If the link nodes are the same node, then a route through level 0 is not required and a route through level 1 is constructed (lines 19–21).

Assuming that a route through level 0 is required, the route is constructed (lines 24–26). Finally, the 3 sections of the route are amalgamated in a *MultiRoute* object which is then returned.

```
1      private static Route buildRoute(long start, long end) {
2    try {
3        //Build A route from <start> to <end>
4        RouterNode startNode = levels[1].getNode(start);
5        RouterNode endNode = levels[1].getNode(end);
6        if ((startNode == null)||(endNode == null))
7          return null;
8        //Find first section in level 1
9        Route firstLeg = new Route(levels[1],startNode.getId(),0);
10       firstLeg.buildRoute(new DijkstraFindUpLink());
11       RouterNode startLink = firstLeg.getFinish();
12       //Find last section in level 1
13       Route endLeg = new Route(levels[1],endNode.getId(),0);
14       endLeg.buildRoute(new DijkstraFindUpLink());
15       RouterNode endLink = endLeg.getFinish();
16
17       if (startLink.getId() == endLink.getId()){
18         //Start and end are directly connected
19         Route testRoute = new Route(levels[1],startNode.getId(),
     endNode.getId());
20         testRoute.buildRoute(new AStar());
21         return testRoute;
22       }
23
24       //Find middle section of route
25       Route middle = new Route(levels[0],startLink.getId(),endLink.
     getId());
26       middle.buildRoute(new Dijkstra());
27
28       //construct multi-route object
29       MultiRoute result = new MultiRoute();
30       result.append(firstLeg,true);
31       result.append(middle,true);
32       result.append(endLeg,false);//don't reverse end leg
33       return result;
34   }catch(Exception e) {
35       System.out.println("Routing failed!");
36       return null;
37   }
38 }
```

Listing 8.9   Building a route using a 2-tier hierarchy

Figure 8.11 shows a 2-tier route from the City of Edinburgh to the City of Aberdeen. We note that the start and end legs traverse local roads to the nearest Motorway/Trunk node; the main leg then uses the motorway/trunk road network for most of the journey.

The sizes of the graphs created when building the example in Fig. 8.11b may be seen in Table 8.5. Note that layer 0 only contains approximately 5% of the total

Fig. 8.11 An example of a
2-tier route from Edinburgh
to Aberdeen

(a) The start leg (level 1).

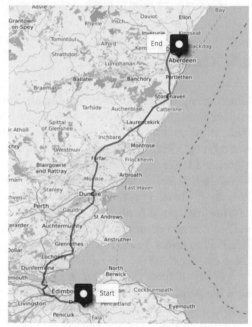

(b) The middle leg (level 0).

(c) The end leg (level 1).

Table 8.5 The graph sizes of the file roads.osm which contains the road network for the whole of Scotland. The sizes for layer 0 (Motorways and Trunk roads) and layer 1 (all other roads) may be seen

	Nodes	Ways
Layer 0	41864	4390
Layer 1	788714	123400
All	828396	127776

nodes, thus the majority of the route comprises a route being found through a much smaller graph. The graph for layer 1 is larger (95%) of the total nodes, but layer 1 is not fully connected and in practice comprises a set of smaller unconnected networks.

8.4   Conclusions

Graph theory and the associated routing algorithms form a foundation for most of the problems examined in this book. A great deal of effort has been expended on ensuring that routing engines such as GraphHopper GmbH (2021) or services such as Google Maps Google Maps (2021) can provide optimised routes in very short times. Such algorithms make use of techniques such as caching, highway hierarchies, contraction hierarchies as well as being implemented in an optimal manner. The examples given here are deliberately simplified in order to make them understandable to the reader.

The principle algorithms examined are Dijkstra, which when using the flood variant carries out an exhaustive search over an entire graph and A* which uses a heuristic to direct the search. There exist many variants based on the principles of these two approaches.

Neither of these approaches scales up especially well to larger networks and so a hierarchical approach that divides the road network into layers can assist in dealing with longer distance routing problems. Hierarchical approaches transform the problem from a search of one large graph to a search of a series of smaller interconnected graphs. Depending on the implementation used, different algorithms may be used at differing layers; also some layers may be searched concurrently and in certain situations routes across layers may be cached to avoid repeated searches for the same route. Caching is an especially powerful technique when solving problems using evolutionary methods as the same route may be requested countless times when a fitness function is evaluating similar individuals. Caching does have a drawback when the problems are dynamic (e.g. taking into account traffic congestion) and cached routes may have a limited life before needing to be re-calculated.

References

Dijkstra, E. W. 1959. A note on two problems in connexion with graphs. *Numerische Mathematik* **1**(1), 269–271. ISBN: 0945-3245. https://doi.org/10.1007/BF01386390

Geisberger, R., P. Sanders, D. Schultes, and D. Delling. 2008. Contraction hierarchies: Faster and simpler hierarchical routing in road networks. In *Experimental Algorithms*, ed. C.C. McGeoch, 319–333. Heidelberg: Springer.

GmbH, G. 2021. GraphHopper Directions API. github.com/graphhopper/graphhopper

Google Maps. 2021. https://www.google.com/maps/

Hart, P.E., N.J. Nilsson, and B. Raphael. 1968. A formal basis for the heuristic determination of minimum cost paths. *IEEE Transactions on Systems Science and Cybernetics* 4 (2): 100–107.

OpenStreetMap contributors. 2021. *Planet dump* retrieved from https://planet.osm.org. Published: https://www.openstreetmap.org

Pavie, Λ. 2021. BasicOSMParser. https://github.com/PanierAvide/BasicOSMParser

Sanders, P., and D. Schultes. 2006. Engineering highway hierarchies. In *Algorithms—ESA 2006*, ed. Y. Azar and T. Erlebach, 804–816. Heidelberg: Springer.

Weber, M. 2020. Osmfilter. https://gitlab.com/osm-c-tools/osmctools

Chapter 9
Linking to Real-world Data Sources and Services

Abstract If we set out to solve real-world problem instances, then we must have the ability to access geospatial data that represents the real world. Such data allows us to produce accurate routes (and therefore journey times and distances); we also need to be able to *geolocate* addresses and place names to accurate coordinates. We commence this chapter by discussing geocoding first by constructing a simple geocoder in order to understand the principles involved and then by making use of the Nominatim service. We then discuss the use of a routing engine, in this case GraphHopper, and the practicalities of using it with search-based metaheuristics such as Evolutionary Algorithms. Finally, we examine data formats that may be used to export data from our solvers into GIS systems, in particular the use of Keyhole Markup Language (KML) which may be used to construct map overlays.

9.1 Introduction

If we wish to construct applications that solve real-world problems, it is necessary to make use of real-world data; we have already discussed the nature of geographical data (Chap. 7) and routing algorithms (Chap. 8). This chapter examines the techniques required to utilise geocoding, mapping and routing services and data within our applications. In many cases, there exists a choice between downloading raw data locally or accessing data online via an API. There exist a growing number of Open Source and commercial GIS tools and data sources which are available to the developer for integration into their problem solver.

Besides integration with data sources, we must also consider the manner in which we present our results to the end user. Raw results might suffice for experimental purposes when developing and evaluating an algorithm, but if we are to construct applications that are of use in real-world scenarios then we must consider how we present our results to the user. We examine how we can translate our results into forms that can be overlayed onto map data or uploaded onto GPS/SatNav devices.

© Springer Nature Switzerland AG 2022

N. Urquhart, *Nature Inspired Optimisation for Delivery Problems*,

Natural Computing Series, https://doi.org/10.1007/978-3-030-98108-2_9

9.2   Geocoding

Geocoding (sometimes referred to as *Address Geocoding*) is the transformation of the name or description of a place into a set of coordinates that specify a precise location. Geocoding provides the link between the textual descriptions that we use to describe the world (addresses, postal codes, etc.) and the coordinates required when representing such locations within the algorithms and systems described in this book.

For example, if someone were to ask me where I work, I could give the answer in one of 2 ways:
Answer 1: Edinburgh Napier University, 10 Colinton Rd, Edinburgh
Answer 2: 55.932941, −3.213922

Both of these answers are correct, but the first is more understandable to another human, the second is best understood if we have a means of plotting latitude and longitude coordinates on a map (see Sect. 9.4). There are a number of other differences; the first answer refers to a University Campus that occupies a large area, and houses several hundred staff and students; the second refers to a specific location of $10 \, cm^2$. If we wish to carry out any distance calculations or routing, then we need to be able to express ourselves in numerical coordinates (typically latitude and longitude). We use numerical coordinates as an internal representation for locations and textual labels or map points when communicating with users.

We need a mechanism with the ability to map between human-readable labels and a formal coordinates system. An intermediate step towards that goal in many countries was the introduction of postal or zip codes allocated to addresses. For clarity within this book, I shall refer to just postal codes rather than postal and zip codes. For example, within the United Kingdom every postal address is associated with an alpha-numeric postcode. Each postcode covers a small number of adjacent addresses within a geographical area. Large businesses or organisations may have their own unique postcode where their volume of mail justifies it.

This gives a third possible answer to the where I work question:
Answer 3: EH10 5DT
(This is the postcode of Edinburgh Napier University)

Every postcode has a point associated with it which, in urban areas, will most likely be close to the addresses associated with that postcode. If I can navigate to the point associated with the postcode, then I should be very close to the actual address (in practice, I will often be able to see the desired house or business from that point). The postal code is a useful intermediate representation of an address; it is more succinct than the full address but easier to memorise and less prone to transcription errors than a set of coordinates. In many countries, the postal code is part of a formal address and so most individuals know their own postcode and are used to treating postcodes as part of an address.

For the UK at least, publicly available data contains the mapping between postcodes and their associated latitude and longitude coordinates; this can be used to build a simple Geocoding system. But there are a number of drawbacks; postcodes can cover several addresses; the coordinates associated with them will represent a central point within the area covered and not necessarily the actual address.

Some countries including Australia, Austria and the Netherlands have made their postal coding data freely available. The reader is referred to Foundation (2021) for a list of how countries license their data.

As we shall see within this chapter, Geocoding services are available which can parse textual addresses and locate them accurately. Such geocoding services require large amounts of data and the ability to parse address strings. For that reason where Geocoding is required, most developers connect to a hosted service allowing them to bypass much of the complexity associated with Geocoding.

Another form of handling coordinates is that of *Geohashing*, which translates coordinate points from numeric data into an alpha-numeric string. For instance, we can use the geohashing service at http://geohash.org to convert the location (55.932941, −3.213922) to the string *gcvwny78ypyj*.

One reason why we might wish to do this is memory use. In Java, a set of lat/lon coordinates would be stored as 2 doubles which requires a total of 16 bytes of memory, whilst the string *gcvwny78ypyj* requires only 12 bytes (assuming UTF-8 string encoding). Whilst this difference of 4 bytes may seem trivial, it could become significant when storing thousands or millions of points within a larger scale system. For instance, there are approximately 30 million addresses in the UK; the difference between storing them all as lat/lon double precision coordinates and as 12 character geohash strings would be in the region of 120 Gb.

Implementing A Simple Geocoder

Geocoding is comparatively simple in concept; in our very simple definition of a geocoder, we only need two methods:

- *geocode()* to convert from a label to coordinates.
- *reverseGeocode()* to convert from coordinates to a label.

Listing 9.1 describes a Java interface for a simple Geocoding service containing the above two methods.

```
1  public interface Geocoder {
2      public LatLon geocode(String label);
3      //Convert a String label to lat/lon coordinate
4      public String reverseGeocode(LatLon p);
5      //Convert a lat/lon coordinate to a String label
6  }
```

Listing 9.1 An interface that specifies a simple Geocoder service

Listing 9.2 gives an implementation of a very basic geocoder called *NapierLocator* which holds data for the locations of the three main campuses of Edinburgh Napier University. In this simple example, we hard code our three locations (lines 8–10), although in a production version we would retrieve this data from a file, allowing it to be easily updated without recompiling. The *geocode* method (lines 14–20) iterates through the list of *Entry* objects to find one that matches the supplied label and returns the associated set of coordinates. An inner class *Entry* (listing 9.3) is used to hold a single mapping of a label to a point.

Reverse geocoding is implemented by the *NapierLoc.reverseGeocode()* method (lines 21–35). At this point, we need to consider that although we geocode a label to a specific point, in practice that label may apply to a larger area than that specific point alone. There are two approaches to resolving this issue:

- Return the label that is closest to the coordinate supplied.
- Return the closest label to the coordinates providing it is within a given threshold.

The first method has the advantage that it will always return a label, but in some circumstances such a label is some distance from the coordinates. In this case, we find the closest entry to the coordinates that we have entered.

```
1    private ArrayList<Entry> data = new ArrayList<Entry>();
2
3    public NapierLocator() {
4        /* Initialise.
5         * In a practical implementation this constructor would load the
     data
6         * in from a CSV file, rather than hard coding
7         */
8        data.add( new Entry(new LatLon(55.932941, -3.213922),"Napier
     University, Merchiston"));
9        data.add( new Entry(new LatLon(55.918029, -3.239706),"Napier
     University, Craiglockhart"));
10       data.add( new Entry(new LatLon(55.924190, -3.288473),"Napier
     University, Sighthill"));
11       }
12
13   @Override
14   public LatLon geocode(String label) {
15       for (Entry e : data) {
16       if (e.label.equals(label))
17           return e.location;
18   }
19   return null;
20   }
21
22   @Override
23   public String reverseGeocode(LatLon p) {
24       double dist = Double.MAX_VALUE;
25       String best = "Not found";
26
27       for (Entry e : data) {
28           double cDist = p.getDist(e.location);
29           if (cDist < dist) {
30               dist = cDist;
31               best = e.getLabel();
32           }
33       }
34       return best;
35   }
```

Listing 9.2 An extract from the *NapierLoc* implementation of *Geocoder*

```
1     private class Entry{
2         /*
3          * An inner class used to store a mapping between a label
4          * and a location (stored as a Visit)
5          */
6         private LatLon location;
7
8         public LatLon getLocation() {
9             return location;
10        }
11
12        public String getLabel() {
13            return label;
14        }
15
16        private String label;
17
18        public Entry(LatLon loc, String label) {
19            this.location =loc;
20            this.label = label;
21        }
22    }
```

Listing 9.3 The inner class *Entry* used within *NapierLoc*

The difficulties in geocoding stem from the amount of data that must be available to the geocoder to look up. In our simple example, both the *geocode()* and *reverseGeocode()* methods have a complexity of θn so they won't scale up as the number of entries in *data* increases. For much larger datasets (e.g. UK postcode data), a hierarchical structure might be more appropriate to speed up the search times.

Using Web-based Geocoding Services

The simple geocoder developed in Sect. 9.2 has many drawbacks, the biggest of which is the sparseness of underlying data. Ideally, we need a geocoder that can accept any valid address string, parse it and return the corresponding coordinates. Connecting to a web-based geocoding service has many advantages when implementing a real-world solver; these include

- **Coverage** Typically, such services will cover large geographical areas, allowing geocoding from a range of labels ranging from country/city/town names to the addresses of individual houses or businesses.
- **Parsing** Address strings can be parsed in multiple ways, in an effort to interpret them correctly. This is particularly useful if the addresses have been entered by the end users themselves.
- **Updating** The burden of keeping the data up to date rests with the service provider and is no longer our responsibility.

Web-based Geocoding services usually comprise an API supported by a comprehensive GIS database. Depending on the accuracy and the type of service required, commercial or free services are available.

Nominatim Nominatim (2021) is a free geocoding service which utilises Open-
StreetMap data. A hosted version of Nominatim is available at https://nominatim.
openstreetmap.org which provides a number of services including geocoding and
reverse geocoding. Listings 9.4 and 9.6 demonstrate an implementation of *Geocoder.
java* that accesses Nominatim. The API is accessed by sending a request (listing 9.4
line 6) over HTTPS (line 7). Our example only uses the most basic features; the Nominatim documentation Overview—Nominatim (2020) describes the capabilities of the
API.

The result returned from a geocoding request may be specified in a variety of
formats, but in this case we specify XML. The XML returned is comprehensive
and contains a great deal of information. For example, querying "Napier University,
Merchiston" returns the XML shown in Fig. 9.1. The XML returned contains much
more than just coordinates; it also contains a full place name, amenity details,
the bounding box rectangle and an OSM place identifier which may be used in
conjunction with OSM data. A more complex implementation could return a data
structure encoding more detail about the information found.

```xml
<?xml version="1.0" encoding="UTF-8" ?>
<searchresults timestamp='Wed, 10 Mar 21 21:33:17 +0000'
   attribution='Data © OpenStreetMap contributors,
   ODbL 1.0. http://www.openstreetmap.org/copyright'
   querystring='Napier University, Merchiston'
   exclude_place_ids='226628788'
   more_url='https://localhost/search/?
   q=Napier+University%2C+Merchiston&amp;
   exclude_place_ids=226628788&amp;format=xml'>

<place place_id='226628788' osm_type='way'
   osm_id='682083727' place_rank='30'
   address_rank='30' boundingbox="55.9323734,
   55.9339043,-3.2157022,-3.2122489" lat=
   '55.93322945' lon='-3.213979159845135'
   display_name='Edinburgh Napier University
   - Merchiston Campus, 10, Colinton Road,
   Merchiston, City of Edinburgh, Scotland,
   EH10 5DT, United Kingdom' class='amenity'
   type='university' importance='0.311'
   icon='https://nominatim.openstreetmap.org
   /ui/mapicons//education_university.p.20.png'/>
</searchresults>
```

Fig. 9.1 The XML returned by Nominatim for the query https://nominatim.openstreetmap.org/
search?q=EdinburghNapier,Merchiston&format=xml

```
1    public LatLon geocode(String label) {
2        LatLon p = null;
3        String request = nominatimBaseURL+ "/search?q="+label+"&format=
     xml";
4        try {
5
6            URL url = new URL(request);
7            HttpsURLConnection connection = (HttpsURLConnection)url.
     openConnection();
8            BufferedReader in = new BufferedReader(new InputStreamReader(
     connection.getInputStream(), "UTF8"));
9
10           String response="";
11           String buffer;
12           while ((buffer =  in.readLine()) != null) {
13               response += buffer;
14           }
15           p = extractLocation(response);
16           in.close();
17       }catch(Exception e) {
18           e.printStackTrace();
19       }
20       return p;
21   }
```

Listing 9.4 The *geocode()* method that connects to Nominatim, note that the *extractLocaton()* method parses the XML returned (see Fig. 9.1) and extracts the location coordinates

```
1        Geocoder gc = new Cache(new Nominatim());
```

Listing 9.5 Using the Cache, note how we use constructors and inheritance to allow the cache to be added with the minimum of changes to the code

```
1    public String reverseGeocode(LatLon p) {
2
3        String request = nominatimBaseURL+ "/reverse?lat="+p.getLat()+ "&
     lon="+p.getLon()+ "&format=xml";
4        try {
5
6            URL url - new URL(request);
7            HttpsURLConnection connection = (HttpsURLConnection)url.
     openConnection();
8            BufferedReader in = new BufferedReader(new InputStreamReader(
     connection.getInputStream(), "UTF8"));
9
10           String response="";
11           String buffer;
12           while ((buffer =  in.readLine()) != null) {
13               response += buffer +"\n";
14           }
15           in.close();
16           return response;
17       }catch(Exception e) {
18           e.printStackTrace();
19           return "";
20       }
21   }
```

Listing 9.6 A very simple implementation of a *reverseGeocode()* method from within a Geocoding service that connects to Nominatim

```
1   public class Cache implements Geocoder{
2       private ArrayList<Entry> cache = new ArrayList<Entry>();
3       private Geocoder baseCoder = null;
4       private final double REVERSE_THRESHOLD = 0.01;
5       //Threashhold for reverse geocoding
6
7       public Cache(Geocoder baseCoder) {
8           /* Any entries not in the cache are resolved using
9            * the baseCoder and then cached */
10          this.baseCoder = baseCoder;
11          }
12
13      @Override
14      public LatLon geocode(String label) {
15          for (Entry e : cache) {
16              if (e.label.equals(label))
17                  return e.location;
18          }
19          //We can assume the label has not been found
20          System.out.println("Adding to cache");
21          LatLon  p = baseCoder.geocode(label);
22          cache.add(new Entry(p,label));
23          return p;
24      }
25
26      @Override
27      public String reverseGeocode(LatLon p) {
28          double dist = Double.MAX_VALUE;
29          String best = "";
30
31          for (Entry e : cache) {
32              double cDist = p.getDist(e.location);
33              if (cDist < dist) {
34                  dist = cDist;
35                  best = e.getLabel();
36              }
37          }
38          if (dist <= REVERSE_THRESHOLD)
39              return best;
40          //Else cache miss
41          String result = baseCoder.reverseGeocode(p);
42          cache.add(new Entry(p,result));
43          return result;
44      }
```

Listing 9.7   A basic cache for geocoding

For those developing large-scale projects which make use of Nominatim (or similar services), hosting your own geocoding service locally is a useful option. For example, Nominatim may be downloaded and set up on your own hardware to provide a dedicated local server. This may not only provide a faster response but also reduce the dependency of your system on third-party services.

Caching Geocoding Requests

We should exercise some consideration when utilising a freely provided web service such as *nominatim.openstreetmap.org*. We must respect the usage limits and fair-use policies of the provider and above all ensure that our usage complies with any terms and conditions associated with use of the service. We face the issue that applies to most web APIs, that of response speed. The time taken to send an HTTPS request and receive the response is significant in comparison to the time taken to query a locally held memory-based data structure. It makes sense, therefore, to restrict our usage of the Nominatim API as much as possible. A simple way of facilitating that is to *cache* the responses. Once a query has been answered by Nominatim, we store the query and the answer in a local data structure. If the query is repeated subsequently, the answer is quickly retrieved from the local cache. This approach of caching can be particularly relevant when using evolutionary-based optimisation techniques where many similar solutions may be evaluated, generating many similar geocoding requests. A word of warning about caching data from web APIs, always check the terms of use attached to the API to ensure that caching is permitted. In some cases, it may not be permitted to cache the results under the terms of the license by which the data is provided.

Listing 9.7 shows a simple implementation of a cache that sits between the client and the Nominatim service. As our Cache class implements *Geocoder*, we can substitute it for any other *Geocoder* as shown in listing 9.5; note that we supply the *Nominatim* object which is used as the source for the cached data to the constructor of the Cache object. The Cache class is a development of the *NapierLocator* class and uses the same inner class *Entry* to represent entries in the cache. The *Geocode()* method (lines 14–24) first searches through data to see if the query has been added to the cache; if the query is not found in the cache, then it is resolved using the *baseCode* object (in this case Nominatim) and added to the cache. The *reverseGeocode()* method (lines 27–44) is more complex. Rather than looking for an exact match of coordinates the closest entry in the cache is found, but only returned if it is within a specified distance threshold (line 38). The value allocated to the threshold (line 4) depends on the application and some experimentation might be required to achieve a reasonable result. The larger the threshold value, the greater the possible distance between the point specified for the reverse geocode and the label returned.

One final point that the reader should consider is if caching data is the time for which that data should be held in the cache. If the cache is made persistent by writing entries saved to a file or database, then the speed advantages of the cache are obtained over multiple sessions. But if the primary data source is liable to be updated, then there will come a point where the cache contains data that has been superseded. There are a number of approaches to dealing with this:

- Not making the cache persistent.
- Deleting the persistence file at a set interval (e.g. weekly or monthly).
- Logging the date that each entry to the cache is made and removing entries when they reach a specific age.

9.3   Using Routing Services

In Chap. 8, we discussed algorithms that may be used for routing. Many of the issues that we discussed in relation to geolocation (see Sect. 9.2) also apply to routing.

For most applications, it makes sense to use a routing engine accessed through an API to find routes between points. As we discovered in Chap. 8, the implementation of even a simple routing engine is a non-trivial task.

There are two possible approaches to utilising a routing engine:

- **Web Hosted** Data and the routing engine are hosted on a server. The client uses the API to request a route, which is constructed on the server and sent to the client.
- **Locally Hosted** The Data and the routing engine are installed on the users' machine. Typically, the routing engine may be a library that can be incorporated directly into the software being developed.

Web-hosting

Arguably, the most well-known of web-based routing services is that offered by Google Maps (2021). Like other services of this nature, it may be accessed by the user through a web interface or by software through an API. As a commercial offering, care needs to be taken to ensure that any use is within the terms of the license associated with the service.

The advantages of linking to such a service include not having the responsibility to maintain and update mapping data. The algorithms used by such services are usually very efficient and flexible including features such as the ability to specify intermediate points. The ability to specify routes by mode (car, walking, public transport, etc.) is also a feature of many services. A number of disadvantages also exist in the adoption of such services. If the service is provided on a commercial basis, then the cost must be considered; also, it may not be permissible to create a cache of routes locally. As with web-based geocoding services, speed of response may also be an issue.

Locally Installed Routing Engines

The use of a locally installed open-source routing engine combined with open-source data presents a flexible solution that allows the user to cache data locally as required and also removes the cost element from routing requests. One of the most commonly used open-source routing engines is GraphHopper GmbH (2021) which uses OpenStreetMap data. GraphHopper is available as a Java library for local use and as a hosted service. Commercial data sources and routing engines are also available in most parts of the world, and the final choice of the data source is down to the requirements that the developer is trying to meet.

There are many similarities between the incorporation of a routing engine into our application and the incorporation of a Geocoding service. We can take advantage of Object-Oriented facilities and specify a *RoutingEngine* abstract class (listing 9.8). This allows us to create concrete instances of *RoutingEngine* that encapsulate different services but are accessed in the same manner, allowing us to change the source of routing data easily without re-writing large amounts of our code. If we are making use of a commercial routing service, this may be especially important as it can allow us to change data providers with the minimum of disruption making it easier to switch data providers.

```
1   public abstract class RoutingEngine {
2       /*
3        * Define some mode keys that can be used within the
4        * options structure
5        */
6       public static String OPTION_DATA_DIR = "dataDir";
7       public static String OPTION_OSM_FILE = "osmFile";
8       public static String OPTION_MODE = "mode";
9
10      public abstract Journey findRoute(LatLon start, LatLon end, HashMap<
        String,String> options);
11      /*
12       * Options holds Key,Value strings that allow options pertaining to a
         specific
13       * RoutingEngine to be set.
14       */
15  }
```

Listing 9.8 The definition of a routing engine as an abstract class

Our requirements for a routing engine are very simple; we only have one method *findRoute()* (line 10) which takes two *LatLon* objects (representing the start and end locations) and some options and returns a *Journey* object. The *HashMap* object *options* is a useful way of allowing a range of *key/value* pairs to be passed to the routing engine; some definitions for keys are given (lines 6–8). One advantage in passing parameters in this way is the ability of the designer to add extra options specific to an implementation without compromising the abstract class.

```
1   public static void main(String[] args) {
2       System.out.println("Homebrew");
3       testRouter(new HomeBrew());
4       System.out.println("GraphHopper");
5       testRouter(new GraphHopper());
6   }
7
8   private static void testRouter(RoutingEngine re) {
9       /*
10       * Set up some options.
11       * Depending in what class re is some options
12       * may not be supported.
13       */
14      HashMap<String,String> options = new HashMap<String,String>();
15      options.put(RoutingEngine.OPTION_OSM_FILE, "roads.osm");
16      options.put(RoutingEngine.OPTION_DATA_DIR, "./data/");
17      options.put(RoutingEngine.OPTION_MODE, "car");
18      Geocoder geocode = new Nominatim();
19      Journey j = re.findRoute(geocode.geocode("Edinburgh"), geocode.
        geocode("Glasgow"), options);
20      System.out.println(j.getPointA() + " : "+ j.getPointB());
21      System.out.println(j.getDistanceKM());
22      System.out.println(j.getTravelTime());
23      for (Visit v: j.getPath())
24          System.out.println(v);
25  }
```

Listing 9.9 Testing the HomeBrew and GraphHopper RoutingEngines

Listing 9.9 demonstrates the use of the *RoutingEngine* class with two concrete implementations. The *HomeBrew* routing engine is the one that was built in Chap. 8. Note the use of *options* to pass parameters to the routing engine and that *HomeBrew* ignores the mode option, as it doesn't support different modes of routing.

Listings 9.10 and 9.11 demonstrate how the detail of the use of a routing API is hidden within the implementation of *RoutingEngine*. The developer may wish to consider implementing a cache in the same manner as for geocoding (see Sect. 9.2), if the license covering data access allows.

```
1    public Journey findRoute(LatLon start, LatLon end, HashMap<String,
     String> options) {
2
3        if (hopper==null)
4            init(options.get(RoutingEngine.OPTION_DATA_DIR),options.get(
     RoutingEngine.OPTION_OSM_FILE));
5
6        GHRequest request = new GHRequest(start.getLat(),start.getLon(),
     end.getLat(),end.getLon()).setVehicle(options.get(RoutingEngine.
     OPTION_MODE));
7        GHResponse response = hopper.route(request);
8        if (response.hasErrors()) {
9            throw new IllegalStateException("S= " + start + "e= " + end +
     ". GraphHopper gave " + response.getErrors().size()
10                      + " errors. First error chained.",
11                      response.getErrors().get(0)
12                      );
13        }
14
15        PathWrapper pw = response.getBest();
16        Journey res = new Journey(start,end);
17        res.setDistanceKM(pw.getDistance()/1000);
18        res.setTravelTimeMS(pw.getTime());
19        ArrayList<LatLon> path = new ArrayList<LatLon>();
20        PointList pl = pw.getPoints();
21        for (int c=0; c < pl.size();c++){
22            path.add(new LatLon(pl.getLatitude(c),pl.getLongitude(c)));
23        }
24        res.setPath(path);
25        return res;
26    }
```

Listing 9.10   An extract from the GraphHopper implementation of the RoutingEngine

```
1    private Graph myGraph;//The OSM Graph
2    private RouterNode startNode= null;
3    private RouterNode endNode= null;
4
5
6    @Override
7    public Journey findRoute(LatLon start, LatLon end, HashMap<String,
         String> options) {
8        if (myGraph == null) {
9            String osmFile = options.get(RoutingEngine.OPTION_DATA_DIR) +
     '/'+options.get(RoutingEngine.OPTION_OSM_FILE);
10           //Build a graph based on the street network in the OSM file
11           myGraph = new Graph(osmFile,null);
12       }
13
14       findNodes(start,end);
15       //Load osm data into a street graph
16       Route testRoute = new Route(myGraph,startNode.getId(),endNode.
     getId());
17       //Create a Route object within the graph based on the start and
     end nodes
18       testRoute.buildRoute(new AStar());
19       //Find a path between the start and the end using A*
20       Journey result = new Journey(start,end);
21       result.setDistanceKM(testRoute.getDist());
22       result.path = new ArrayList<LatLon>();
23
24       for(LatLon loc :testRoute.getLocations()) {
25           result.path.add(0,new LatLon(loc.getLat(),loc.getLon()));
26           //Reverse the order by always adding at 0
27       }
28       return result;
29   }
```

Listing 9.11 An extract from the *HomeBrew* implementation of the *RoutingEngine*

9.4 Exporting Data

To ensure that our system is usable in real-world contexts, it is necessary to ensure that results are presented in formats that are compatible with standards used by other GIS tools. Many useful data formats exist which allow data to be imported into other systems; helpfully, many of these formats are based around ASCII text files, examples include

- **KML** Keyhole Markup Language (KML) OGC (2015): An XML-based format originally defined for use with Google Earth. KML is used to define overlays which may then be displayed on top of maps; overlays comprise components such as polygons, lines, place markers or labels. KML components can be associated with latitude/longitude coordinates which allows the creator to specify exactly where they should be placed in relation to the underlying map. KML can be used as a means of visualising results such as delivery routes. Many products are available that can render KML files over a map; these include web APIs such as Leaflet Vladimir Agafonkin (2020) as well as a number of online KML viewers. KML files can be made available on web servers and displayed in browsers via HTTP.

- **GPX** GPS exchange: An XML-based file format that may be used to import data into GPS or Sat Nav systems. GPX files are structured around a series of *way points*, which collectively form a *route*. Individual way points are connected via *tracks* which comprise a series of latitude/longitude coordinates that specify the route to be taken. GPX files may be especially useful for vehicle routing problems, as they can allow an optimised route to be transferred directly to an in-vehicle Sat Nav system.
- **CSV** Comma Separated Values: A text-based file format commonly used to exchange data between software packages. Data within a CSV file is laid out in a tabular format, with rows in the file corresponding to rows in the table, each row being split into columns by commas. The first row in a CSV file may be treated as a header. A CSV file is a simple way of exchanging data to a wide variety of spreadsheets and database systems.

Exporting from Java

As with Geocoding and Routing services, we adopt an object-oriented approach that allows us to add additional export options with the minimum of re-engineering. Listing 9.12 defines an export service, for exporting routes and way points.

```
1   public interface ExportService {
2       void addTrack(ArrayList<LatLon> locs);
3       //Draw a track along the lost of LatLons provided
4       void addWaypoint(LatLon loc, String caption, String desc);
5       //Add a waypoint at <loc>
6       void write(String path, String name);
7       //Write the contents to a file at <path> using <name>
8   }
```

Listing 9.12 A definition of a simple export service

A concrete implementation of *ExportService* is given in listing 9.13 which demonstrates writing a .GPX file. Note that the GPX code produced is basic, and the reader may wish to modify this class in order to utilise the full range of features supported by GPX files. Listing 9.14 demonstrates the manner in which the *ExportService* may be used in conjunction with the *Geocoder* and Routing services discussed earlier. Viewers are available online for the GPX and KML files which will overlay them onto maps. Figure 9.2 shows the KML generated by listing 9.14 overlaid onto OSM mapping data.

Fig. 9.2 The GPX output of *ExportTest*. Visualised using https://www.gpsvisualizer.com/ *Base map and data from OpenStreetMap and OpenStreetMap Foundation* (https://www.openstreetmap. org/copyright)

```
1   public class GPXWriter implements ExportService  {
2       private  String segments = "",waypoints ="";
3
4       @Override
5       public  void addTrack(ArrayList<LatLon> locs) {
6           for (LatLon l : locs){
7               segments += "<trkpt lat=\"" + l.getLat() + "\" lon=\"" + l.
    getLon() + "\">"+"</trkpt>\n";
8           }
9       }
10      @Override
11      public void addWaypoint(LatLon l,String caption, String desc) {
12          desc = caption +" " + desc;
13          waypoints +=
14                  "<wpt lat=\""+l.getLat()+" \" lon=\" "+l.getLon()+"\">"+"
    <name>"+desc+ "</name></wpt>";
15      }
16      @Override
17      public  void write(String file, String name) {
18          file = file +".gpx";
19          name = "<name>" + name+ "</name>\n";
```

```
1          try {
2              FileWriter writer = new FileWriter(file, false);
3              writer.append(header() +name + waypoints + "<trk><trkseg>" +
       segments + footer());
4              writer.flush();writer.close();
5              waypoints = "";segments = "";
6          } catch (IOException e) {
7              e.printStackTrace();
8          }
9      }
10     private String header() {
11         return "<?xml version=\"1.0\" encoding=\"UTF-8\" standalone=\"no
       \" ?>\n"
12
13                 + "<gpx xmlns=\"http://www.topografix.com/GPX/1/1\" "
14                 + "xmlns:xsi=\"http://www.w3.org/2001/XMLSchema-instance
       \"  xsi:schemaLocation=\"http://www.topografix.com/GPX/1/1 "
15                 + "http://www.topografix.com/GPX/1/1/gpx.xsd\">\n";
16     }
17     private String footer() {
18         return "</trkseg></trk></gpx>";
19     }
20 }
```

Listing 9.13 An implementation of ExportService that exports to .GPX files

```
1      public static void main(String[] args) {
2          testWriter(new ConsoleWriter());
3          testWriter(new KMLWriter());
4          testWriter(new GPXWriter());
5          testWriter(new CSVWriter());
6      }
7
8      private static void testWriter(ExportService export) {
9          //Demonstrate the exporting of data via <export>
10         //1) Create two waypoints
11         Geocoder geocode = new Nominatim();
12         String startAddress = "10 Colinton Road, Edinburgh";
13         LatLon start = geocode.geocode(startAddress);
14         String endAddress = "Scottish Vintage Bus Museum, Fife";
15         LatLon end = geocode.geocode(endAddress);
16
17         //2) Create a route between the waypoints
18         RoutingEngine router = new GraphHopper();
19         HashMap<String,String> options = new HashMap<String,String>();
20         options.put(RoutingEngine.OPTION_OSM_FILE, "roads.osm");
21         options.put(RoutingEngine.OPTION_DATA_DIR, "./data/");
22         options.put(RoutingEngine.OPTION_MODE, "car");
23         Journey route = router.findRoute(start, end, options);
24
25         //3) Export the waypoints and route
26         export.addWaypoint(start, "Start", "Edinburgh Napier University")
       ;
27         export.addWaypoint(end, "Destination ", "Scottish Vintage Bus
       Museum");
28         export.addTrack(route.getPath());
29         export.write("out", "./");
30     }
```

Listing 9.14 A demonstration of the use of ExportService to export data. Full listings for KMLWriter ConsoleWriter and CSVWriter can be found in the code repository that accompanies this book

9.5 Conclusions

This chapter has demonstrated how the developer may integrate real-world data sources and services. Possibly, the most significant development in the last decade within this area has been the development of OpenStreetMap Haklay and Weber (2008). OSM provides an open source of geographical data that may be used by any developer for any purpose (subject to the terms of the OSM licensing agreements). Prior to OSM, access to such data was usually on a commercial basis which put it out of the reach of many developers. Alongside OSM has been a growth in software applications such as Nominatim Nominatim (2021) and GraphHopper GmbH (2021) which allow developers to exploit OSM data with comparative ease. The adoption of formats such as KML and GPX allows us to export results to applications including GIS and Sat Nav software. The effect of all this is to make it far easier for the specialist developer to create an algorithm that optimises a particular problem and then integrates the algorithm into other GIS and business systems and workflows.

References

Foundation, O. K. 2021. Global Open Data Index: Locations. https://index.okfn.org/dataset/postcodes/

GmbH, G. 2021. GraphHopper Directions API. https://github.com/graphhopper/graphhopper

Google Maps. 2021. https://www.google.com/maps/

Haklay, M., and P. Weber. 2008. OpenStreetMap: user-generated street maps. *IEEE Pervasive Computing* 7 (4): 12–18.

Nominatim. 2021. https://nominatim.org/

OGC. 2015. KML 2.3. http://docs.opengeospatial.org/is/12-007r2/12-007r2.html

Overview—Nominatim Documentation. 2020. https://nominatim.org/release-docs/develop/api/Overview/

Vladimir Agafonkin. 2020. Leaflet.

Part III
Real-World Case Studies

This section discusses a selection of real-world urban logistics problems which have been solved using Nature Inspired Techniques. Each of the problems discussed has been investigated by the author in conjunction with academic colleagues and industry partners.

In each case, the software has been re-implemented along the lines suggested in this book and has been made available in the accompanying software repository. Where it has not been possible to distribute the original datasets, similar data has been provided.

Chapter 10
Delivering Food

Abstract This chapter provides a detailed description of the Foodel application. Foodel is a Vehicle Routing Problem (VRP) solver that is designed to be used by small businesses and community organisations to organise and optimise home deliveries. Foodel was developed as a response to the COVID-19 pandemic and was built upon algorithms and techniques outlined earlier in this book. This chapter focuses on the *soft computing* elements of the implementation such as a fitness function that matches the users' expectations, overlaying answers onto maps and having an input format (CSV) that can be easily set up in a spreadsheet.

10.1 Introduction

During 2020 the UK, along with many other countries, entered a period of lockdown due to the COVID-19 pandemic. Within these changes was a requirement for cafes and other food outlets to close to the public, combined with a general instruction for individuals to stay at home as much as possible. This led to a number of organisations and businesses setting up home delivery services rapidly. Businesses which had previously traded as Cafes found themselves providing online delivery services. Using the techniques outlined in this book, the author developed a simple application called Foodel Neil Urquhart (2020), designed to allow the user to quickly solve simple vehicle routing problems and was aimed at those working in response to the COVID-19 pandemic.

The problem solved by Foodel is the Capacitated Vehicle Routing Problem (CVRP) where one or more routes are required in order to allow deliveries to be made to customers. There are two constraints which determine the number of delivery routes:

- The capacity of the vehicle.
- The maximum time that a delivery may be in transit.

Each delivery has a demand associated with it; although this is an abstract numerical value, in this instance it is described as "bags" as in quantity of shopping bags. The vehicle capacity was deemed to be the maximum number of bags that could be carried in the vehicle. In other applications, the demand/capacity was described as

© Springer Nature Switzerland AG 2022
N. Urquhart, *Nature Inspired Optimisation for Delivery Problems*,
Natural Computing Series, https://doi.org/10.1007/978-3-030-98108-2_10

meals or trays depending on what was being delivered. The time in transit constraint refers to the time (in minutes) between setting off from the start and making the final delivery. This time constraint was particularly important when delivering food, as it represented the maximum time that the food order could be left in a vehicle.

10.2   Fitness

Reducing the total distance covered would appear to be the most appropriate fitness function for this problem, given the objective of having the deliveries made as efficiently as possible. As with many real-world problems, feedback from users provides food for thought.

Consider the simple problem shown in Fig. 10.1: 7 deliveries must be made, starting and ending at the point labelled "start". Three solutions a, b and c are presented, each involving travelling the same physical distance. If an EA used length as the fitness function, then there is an equal probability that a, b or c will be evolved. Feedback from users indicated that they considered b and c to be inferior to a. The difference between a, b and c is b and c encode a route that involves passing some delivery points prior to delivering to them later in the route. In solution b, the user travels all of the way to the furthest point passing each delivery point, making the

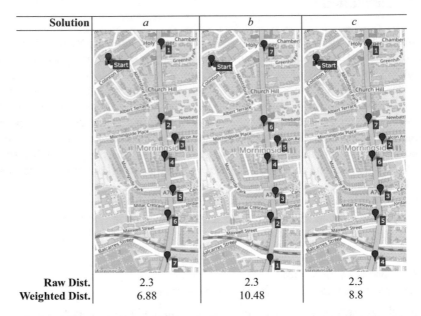

Solution	a	b	c
Raw Dist.	2.3	2.3	2.3
Weighted Dist.	6.88	10.48	8.8

Fig. 10.1 In this simple example, 7 deliveries must be made; they are all located along the same road. Each of these 3 possible solutions has the same travel distance. As described in the text, users will most likely perceive a to be a more desirable solution than b or c. *Base map and data from OpenStreetMap and OpenStreetMap Foundation* (https://www.openstreetmap.org/copyright)

deliveries whilst returning. Answer c makes some deliveries whilst travelling to the furthest point and some on returning. In many cases, the time constraint was set to a level (e.g. 2 h) that meant that solutions such as b and c would not invalidate it. Feedback from users suggested that whenever a route took them past a delivery, they expected to make the delivery and not at a later stage in the route. We can use a weighted distance metric as the fitness function which will reward those solutions that undertake deliveries as soon as possible within the round.

The weighted distance multiplies the distance of each stage of the route by the number of deliveries still to be made. Suppose we have deliveries A–E to make, starting and finishing from a depot:

Distance		2	5	6	5	3	4	=	25
Route	Depot	A	B	C	D	E	Depot		

The fitness of the above route would be 25 based on the total distance travelled.

Distance		2	5	6	5	3	4	=	25
WC[1]		6	5	4	3	2	1		
WD[2]		12	25	24	15	6	4	=	86
Route	Depot	A	B	C	D	E	Depot		

The weighted distance is 86. To minimise the weighted distance, we need to place the shortest length journey segments as early as possible in the route. This favours undertaking as many deliveries at the earliest stage possible within the route (Fig. 10.1).

10.3 Implementation Issues

Our implementation makes use of some of the classes developed in previous chapters. The source code may be found in the repository that accompanies this book.

Figure 10.3 shows the structure of our demonstration application. At first glance, the class diagram may look a little complex, but it can quickly be broken down into three layers:

- The classes which have been imported from earlier chapters which provide us with basic data structures and algorithms.
- The classes created for this application which specialise the evaluation function, operators and representations (in this case, these classes all have names beginning with *Food*).
- The application class which accesses the optimisation functionality through a *Facade* pattern Erich Gamma et al. (1994) named *FoodFacade*.

The use of the *Facade* pattern allows us to keep the implementation of the optimisation separate from the rest of the application. The application may be part of a larger system that incorporates functionality such as online ordering, stock control or payment processing.

Listings 10.1 and 10.2 show the application code that loads the problem instance from a CSV file (see Table 10.2 for a description of the CSV structure used).

```
1  public class FoodDeliveryApp {
2    public static void main(String[] args) {
3      //Load a problem from a file
4      FoodFacade food = loadProblem(new File("./problem.csv"));
5      //solve the problem
6      food.solve();
7      //save the solution in a variety of formats
8      try {
9        food.save(SaveTo.CONSOLE);
10       food.save(SaveTo.KML);
11       food.save(SaveTo.GPX);
12       food.save(SaveTo.CSV);
13     }catch(Exception e) {
14       e.printStackTrace();
15     }
16   }
```

Listing 10.1 The food application

The FoodFacade class provides an API for the developer of the main application to make use of; Table 10.1 shows the methods that have been exposed to the application programmer. The application can specify a problem, and request that it be solved and have the solution saved in a specific format. Development of the API might involve returning the solution as a data structure into the application. Details such as the optimisation algorithm, pathfinding and geocoding are kept hidden from the applications programmer who could incorporate the solver within a desktop application or a web service as required.

When developing an optimisation algorithm that is to be packaged into a larger application is necessary to store parameters such as mutation rates or population

Table 10.1 The methods exposed by the FoodelFacade

Method Signature	Notes
public FoodFacade()	Constructor
public void solve()	
public void addVisit(String name, String address, int demand, String order)	Add a visit
public void setStart(String addr)	
public void setReference(String ref)	
public void save(SaveTo save)	
public void setStartDateTime(String dateTime)	
public void setRoundMode(String type)	
public void setCapacity(int cap)	
public void setMinsDel(int minsDel)	
public void setMaxMinsRound(int maxMinsRound)	

sizes. Such parameters present a quandary; they may need to be altered to improve the performance of the solver in certain situations, but if they are altered by someone who does not understand their effects, then the result may be poor performance. One approach is to hard code the values into the source code; this has the advantage of restricting access only to those who possess the source code, but the disadvantage of requiring recompilation each time a parameter is modified (also it cannot be denied that hard coding such values into the source code is regarded as very poor software engineering practice). Java provides a very useful solution to this problem in the form of a properties file, a text file containing a set of key, value pairs, which can be easily loaded into a Properties object. Figure 10.2 gives an example of the properties file used with the food application; the properties are loaded and managed by the *FoodProperties* class (see Fig. 10.3).

Using the techniques discussed in Chap. 9, we use Nominatim for geocoding address strings and GraphHopper for pathfinding. We also implement a simple cache, to avoid repeatedly running GraphHopper to produce the same paths.

In a real-world context, the speed of solving the problem becomes an issue. When undertaking experimental work (as in earlier chapters), it might be appropriate to utilise a fixed evaluations budget, but a user is likely only to be willing to wait a finite time for a solution. We limit our algorithm by time (as specified within the properties file), no matter what size of problem or specification of hardware the software is running upon then a solution will be produced within a maximum time. The time to be allowed depends on a number of factors, not least the context of the problem, an example is a solution required ASAP or are we scheduling for the next day? Also to be taken into account are the users' patience and the hardware specification that the system is running on. The value to be chosen for the execution time is essentially a

```
#The properties associated with the Food Delivery Algorithm
xo_rate=0.05
population_size=50
#Pool size for tournament selection
tour_size=2
#Run time in seconds (per run)
runtime=120
#No of times to repeat
repeat=5
#Mutation probabilities % that a mutation will be nearest neighbour
or inverted.
#These two values normally add up to <1, as they are alternatives to the
default (random) mutation
mutnn=0.4
mutinvert=0.4
#Use the weighted distance (distance * remaining visits)
weightdistance=true
#OSM File
osmfile=scotland-latest.osm.pbf
datadir=./data/
```

Fig. 10.2 An example properties file as used in the food application

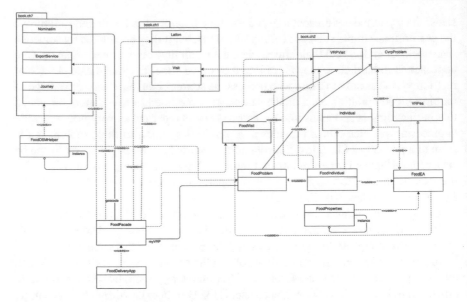

Fig. 10.3 The classes used within the food routing application

matter for the developer to decide in conjunction with the user. Our strategy is simply to allow a fixed time, but other possible strategies include

- **Stop and Resume**
  We present the user with a GUI that has "Pause" and "Accept" buttons. As the algorithm executes, details of the best solution found so far are displayed on the screen. At any time, the user can use the pause button to halt the algorithm, possibly allowing them to view the current best solution in more detail. If the solution is not what is required, the user can press pause once more to resume evolution. When a solution has been found that the user is satisfied with, they can use the "Accept" button to have that solution accepted. This approach could be combined with multiple runs—if the user has not accepted a solution within a given time, restart the evolution with a new random population. This approach requires a more complex GUI and in particular will probably require the solver to be executed on a separate thread to the GUI.
- **Timeout**
  Waiting until a specified number of evaluations has taken place without any improvement to the best solution, but with a timeout constraint. This should allow the algorithm to evolve a useful solution. This approach can also be combined with multiple runs. The use of an overall time limit ensures that the user will receive an answer within a guaranteed time.

The strategy chosen by the developer depends very much on the context of the problem and the type of user.

```
1  private static FoodFacade loadProblem(File file) {
2    //Return a new FoodFacade based on the problem within the
     file
3    try {
4      FoodFacade food = new FoodFacade();
5      Scanner myReader = new Scanner(file);
6      while (myReader.hasNextLine()) {
7        processLine(food, myReader.nextLine());
8      }
9      myReader.close();
10     return food;
11   } catch (FileNotFoundException e) {
12     System.out.println("An error occurred.");
13     e.printStackTrace();
14     return null;
15   }
16 }
17
18 private static void processLine(FoodFacade hf,String line) {
19   //Process a single line of the .CSV file
20   String[] data = line.split(",");
21   data[0] = data[0].toLowerCase().trim();
22   if (data[0].equals("ref")) {
23     hf.setReference(data[1]);
24   }
25   if (data[0].equals("commences")) {
26     hf.setStartDateTime(data[1]);
27   }
28   if (data[0].equals("start")) {
29     hf.setStart(data[1]);
30   }
31   if (data[0].equals("visit")) {
32     hf.addVisit(data[2],data[3],Integer.parseInt(data[4]),""
     );
33   }
34   if (data[0].equals("capacity")) {
35     hf.setCapacity(Integer.parseInt(data[1]));
36   }
37   if (data[0].equals("mins/delivery")) {
38     hf.setMinsDel(Integer.parseInt(data[1]));
39   }
40   if (data[0].equals("time/round")) {
41     hf.setMaxMinsRound(Integer.parseInt(data[1]));
42   }
43   if (data[0].equals("mode")) {
44     hf.setRoundMode(data[1]);
45   }
46 }
47 }
```

Listing 10.2 The food application problem loading and parsing methods (see Fig. 10.2)

Whilst considering implementation issues, it is worth giving some thought to the format of the problem input file. The parsing of input data has deliberately not been encompassed within *FoodFacade*. The source of the data could range from a file created by the user and uploaded, to data (e.g. JSON) received from another application such as an e-commerce platform. In the case of our application, we adopt a simple .CSV file that contains the problem instance. Discussions with potential

Table 10.2 The input data format adopted. The underlying format is .CSV

	A	B	C	D	E
1	Ref	Edinburgh Test Problem			
2	Capacity	30			
3	mins/delivery	5	mins		
4	time/round	500	mins		
5	Mode	car			
6	commences	04/05/2021 09:00			
7	Start	Edinburgh Napier University 10 Colinton Rd Edinburgh EH10 5DT			
8	Visit	1	National Museum of Scotland	Chambers St. Edinburgh EH1 1JF	1
9	Visit	2	Museum on the Mound	The Mound. Edinburgh EH1 1YZ	1
10	Visit	3	People's Story	Royal Mile Edinburgh EH8 8BN Scotland	1
11	...				

users suggested that the one file format which most users could manage to edit (usually using a spreadsheet) was .CSV. The format adopted uses column A to contain a keyword indicating what type of data is contained on that row, and subsequent columns to contain values relating to that keyword. Any row with an empty column A or a value that is not a keyword is skipped. An example of this simple format may be seen in Table 10.2; the simple parsing code is given in Listing 10.2.

10.4 Results

The problem instances that were solved were not in themselves computationally significant; many had less than 20 deliveries per instance, the largest just under 100. The challenge was gaining the trust of the users. The original problem instances used were confidential to the users themselves and cannot be made available; instead, a sample problem instance has been created, based on the addresses of popular Edinburgh tourist attractions (Fig. 10.4); we'll refer to this as the ETA instance.

Tables 10.3 and 10.4 show the effects of increasing the vehicle capacity and transit time constraints. As might be expected as the problem becomes more constrained, the number of delivery routes increases.

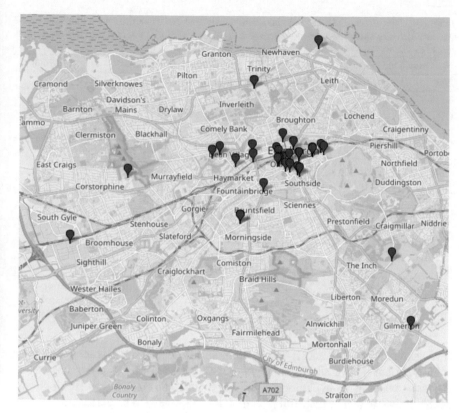

Fig. 10.4 Edinburgh tourist attractions. *Base map and data from OpenStreetMap and Open-StreetMap Foundation* (https://www.openstreetmap.org/copyright)

Table 10.3 The effect of altering the delivery capacity constraint for the ETA instance. Note that capacity is best used where it is a multiple of the total demand (30)

Delivery capacity	Fitness	Routes	Tot time	Avg. route time	Avg. dels/route
4	129.50	8	441.55	55.19	3.75
8	141.12	4	397.59	99.40	7.50
10	144.22	3	386.31	128.77	10.00
12	142.79	3	387.22	129.07	10.00
14	144.72	3	392.45	130.82	10.00
16	175.07	2	378.75	189.38	15.00
18	165.37	2	377.64	188.82	15.00
20	175.03	2	376.99	188.49	15.00
22	170.18	2	376.99	188.49	15.00
24	178.51	2	378.75	189.38	15.00
26	183.94	2	378.36	189.18	15.00
28	197.57	2	381.19	190.59	15.00
30	248.40	1	367.47	367.47	30.00

Table 10.4 The effect of altering the maximum transit time. Note that the transit time is the time that each delivery spends in transit—where that is as low as 20 min, the average delivery time is greater as that includes the time to return to the depot *after* the last delivery

Max. transit time	Fitness	Routes	Tot time	Avg. time	Avg. dels/route
20	169.21	27	688.84	25.51	1.11
40	130.59	11	483.88	43.99	2.73
60	127.62	8	443.77	55.47	3.75
80	126.52	8	443.50	55.44	3.75
100	127.40	8	442.08	55.26	3.75
120	129.90	8	444.89	55.61	3.75
140	127.54	8	445.16	55.64	3.75
160	128.48	8	445.80	55.73	3.75
180	129.64	8	446.41	55.80	3.75
200	127.46	8	442.77	55.35	3.75
220	127.97	8	444.03	55.50	3.75
240	127.42	8	443.36	55.42	3.75

10.5   Conclusions

This chapter is not so much about optimisation techniques, which have been extensively covered in earlier chapters, but about some of the *softer* issues that may arise when using evolutionary (or other stochastic) techniques. If stochastic techniques are to realise their full potential, then they have to be presented in a way that allows non-experts to trust them and also to allow software developers who are not necessarily optimisation specialists to easily integrate them into supply chain or e-commerce systems.

Based on the experience with Foodel and other applications, a number of conclusions may be drawn which may ease the adoption of evolutionary techniques:

1. **Package**: Place the optimisation algorithm within an archive such as a .JAR file (for Java code) to allow for easy distribution.
2. **API**: Use a mechanism such as a Facade pattern to implement an API giving access to the optimiser. Only expose a set of high-level methods that allow a problem to be specified and the answer obtained. Keep the implementation of specialist algorithms and techniques for functions such as optimisation, geocoding and pathfinding "behind" the API.
3. **Configuration File**: Place parameters such as population size of mutation rate into a text file (e.g. a Java properties file). Restrictions could be placed on allowing users to modify such a file to prevent unauthorised tampering, but using a text file allows an expert to adjust the algorithm parameters without recompiling the software. If the algorithm is likely to be used in a variety of scenarios, having customised properties file for each scenario allows the algorithm parameters to be optimised for each scenario whilst maintaining a common code base.

4. **Manage Execution Time**: Users may often be sensitive to the time taken to find a highly optimised solution. It has been the experience of the author that some users are apt to accept a very good (but sub-optimal) solution if it is produced within a short timescale (see Fig. 3.3b). Besides simply halting the evolutionary loop after a fixed time, other strategies such as Stop and Resume or Timeout (see Sect. 10.3) could be implemented depending on the manner in which the software is being used. One option may be to allow different strategies to be selected based on an entry in the configuration file.

5. **Multiple Runs**: When using a stochastic algorithm, it is desirable to execute the algorithm multiple times to increase the probability of finding a higher quality solution. Such multiple runs can use a significant amount of time. Reducing the number of runs may be a useful way of allowing the user the option of trading solution time against solution quality (i.e. a solution can be produced more quickly, but there is a greater chance of it being of lesser quality).

6. **Fitness Function**: It is imperative that the criterion used by the fitness function to identify high-quality solutions aligns with the users' definition of a high-quality solution. For instance, if a user judges the quality of a delivery route by the time taken to make deliveries rather than the distance covered, the fitness function should be time- rather than distance-based. In the Foodel example, it was identified that a simple distance calculation did not produce results that matched the users' expectations; see Sect. 10.2.

It may be argued that the key concept is that of *trust* Andras et al. (2018); Guckert et al. (2021); we must ensure that we present our techniques in a way that builds trust in them.

Acknowledgements The work in this section was encouraged by Joanna and Daniel Campbell the owners of the Leaf and Bean Cafe in Comiston Road, Edinburgh, Scotland. Along with many other organisations, large and small, they rose to the challenge of supplying sustenance directly to the homes of individuals during the COVID-19 lockdown. Their support, encouragement and feedback were immeasurably useful when developing the techniques described in this chapter. Feedback and encouragement were also received from Edinburgh Community Food and Edinburgh Food Social.

References

Andras, P., L. Esterle, M. Guckert, T. A. Han, P. R. Lewis, K. Milanovic, T. R. Payne, C. Perret, J. Pitt, S. T. Powers, N. Urquhart, and S. Wells. 2018. Trusting Intelligent Machines: Deepening Trust Within Socio-Technical Systems. *IEEE Technology and Society Magazine* 37 (4): 76–83. https://doi.org/10.1109/MTS.2018.2876107.

Gamma, Erich, Richard Helm, Ralph Johnson and John Vlissides. 1994. *Design Patterns: Elements of Reusable Object-Oriented Software*. Addison Wesley.

Guckert, M., N. Gumpfer, J. Hannig, T. Keller, and N. Urquhart. 2021. A conceptual framework for establishing trust in real world intelligent systems. *Cognitive Systems Research* 68: 143–155. https://doi.org/10.1016/j.cogsys.2021.04.001.

Neil Urquhart. 2020. Foodel : A Simple Tool for Organising Home Deliveries. www.foodel.info.

Chapter 11
Delivering Letters

Abstract The delivery of items to households over a wide area where each household receives a delivery/visit is a well-known problem. The most common instance of this problem is the letter/parcels service offered by most state-run postal carriers (e.g. The Royal Mail in the UK); in such scenarios, the operative walks through the streets visiting houses. Other instances of this problem include political canvassers or census takers. We can represent this problem as a Travelling Salesperson Problem (TSP), but the TSP does not scale well (see Chap. 1). This chapter introduces a representation and decoder known as *Street Based Routing* (SBR) which uses the groupings of houses by street to reduce the size of the chromosome. We show that an Evolutionary Algorithm (EA) combined with SBR can produce a solution far quicker than the equivalent EA.

11.1 Introduction

Some problems may benefit from a representation and operators that are tailored towards that problem. This chapter examines the problem of house-to-house deliveries in urban areas. The work discussed in this chapter was previously described in Urquhart et al. (2001).

Postal workers, charity collectors and others may need to find the shortest route when calling every house in an area. We can formulate this as a Travelling Salesperson Problem, but as we recall from Chap. 1, solution space for a tsp is $n!$. As we are routing workers who can walk at the same speed in both directions, we can modify the solution space size to $\frac{n!}{2}$. Even with this reduction in size, we still face the problem of large solution spaces for relatively small problems.

Within an urban area, individual addresses can be grouped together into streets, streets can be broken up into sections that join junctions and individual addresses can be further grouped by street sides. Figure 11.1 shows an extract from a map, where sets of addresses have been grouped together; the groups correspond to two criteria:

© Springer Nature Switzerland AG 2022
N. Urquhart, *Nature Inspired Optimisation for Delivery Problems*,
Natural Computing Series, https://doi.org/10.1007/978-3-030-98108-2_11

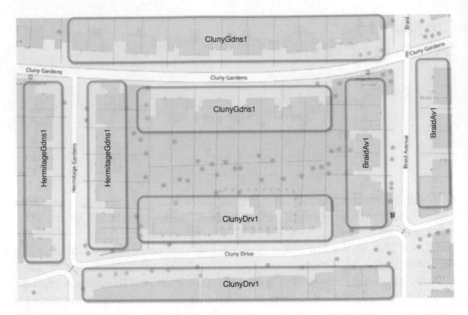

Fig. 11.1   Four urban streets, with the houses grouped by street section and side. *Base map and data from OpenStreetMap and OpenStreetMap Foundation* (https://www.openstreetmap.org/copyright)

- All of the buildings are on the same side of the road.
- All of the buildings are located between two junctions.

Figure 11.1 shows how 61 individual addresses can be grouped into 8 clusters. Each street section may be associated with one or two junctions. The majority of sections link two junctions, but some streets (i.e. cul-de-sacs or dead ends) only have one junction linking them to the rest of the network. If we treat the 61 addresses as a permutation, the number of possible permutations is $\frac{n!}{2}$ or 2.5×10^{83}. As we might recall from Chap. 1, searching spaces of this magnitude is a non-trivial task. It can be argued that some of the solutions representing the shortest tours will keep the houses within the groups shown in Fig. 11.1; if an address is located in the middle of a street section, then it is most likely that it will be best placed in a tour alongside those addresses that are physically adjacent to it. If we accept that we will keep the addresses within their clusters, the number of possible permutations is $\frac{8!}{2}$ or 20,160. If our genotype is a permutation of street sections, then there will be far fewer permutations to search through. In this chapter, we discuss a representation that allows us to have a genotype of street sections that decodes to a phenotype of house addresses, representing the route to be followed.

Table 11.1 A valid genotype based on the sections shown in Fig. 11.1

ClunyGdns1
ClunyDrv1
ClunyDrv1
BraidAv1
HermitageGdns1
HermitageGdns1
BraidAv1
ClunyGdns1

11.2 Street Based Routing

Representation

An example of a valid genotype, of street sections, for the streets shown in Fig. 11.1 is shown in Table 11.1. When decoding the phenotype, each gene is expanded into its set of individual addresses. In our example, the gene "BraidAv1" would be expanded into addresses 2, 4, 6 and 10 Braid Avenue. We need to determine the ordering in which the individual addresses should be added to the phenotype for which we need to know three things:

1. The end of the section (junction) that we start from.
2. The end of the section (junction) that we are to end up at.
3. Whether we are delivering to one or both sides in this operation.

Item 1 can be determined by locating which of the street section junctions is closest to the previous delivery, and item 2 is determined by which of the two junctions is closest to the next street section in the genotype. Where a section has two sides, the gene appears twice in the chromosome, once for each side. If the two genes for a section are adjacent, then they are interpreted as delivering to both sides in one operation (e.g. ClunyDrv1 in the example chromosome); if they are separate, then the sides are delivered separately at different points in the route (e.g. ClunyGdns in the example). Examples of patterns may be seen in Table 11.2.

A useful feature of this scheme is the ability to pre-process each street section. By using the nearest neighbour algorithm, orderings for each side and for the combined sides can be determined at startup. Having the delivery orders and associated distances cached allows us to reduce the amount of processing required during each evaluation.

Decoding from the Genotype to the Phenotype

The basic genotype is a collection of street sections as described in Sect. 11.1; a complication of the SBR representation is the existence of duplicate genes in the chromosome where a street section is double-sided.

Table 11.2 Examples of the delivery patterns; the choice of pattern is based upon whether the delivery operative needs to deliver to both sides at this point and which junction they need to end up at in order to be best placed to commence the next street section

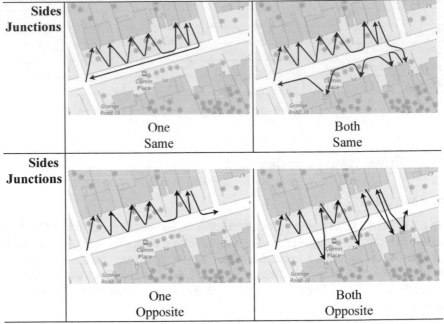

Sides			
Junctions			
One		Both	
Same		Same	
Sides			
Junctions			
One		Both	
Opposite		Opposite	

The main loop of the decoder is shown in Algorithm 22; this iterates through the genotype (line 6), determining whether each street section is to be treated as a double- or single-sided delivery (lines 7–9) at that point. The *applyPattern()* function (Algorithm 23 is used to determine which of the delivery patterns (Table 11.2) should be applied. In order to apply the pattern, it is necessary to calculate which junction is closest to the previous street; this becomes the starting point for the current street section (line 2). The end junction is the junction closest to the next street (line 3).

If the start and end junctions are the same (lines 5–6), then a pattern should be applied that allows the delivery person to end up at the same junction as they started from, else the pattern should leave them at the opposite end of the street. Lines 7–19 select and apply the appropriate pattern, calculating the distance walked, including ensuring that the delivery person ends up at the correct junction for walking to the next street. The returned value *deliveries* is a collection containing the individual deliveries in correct order (Fig. 11.2).

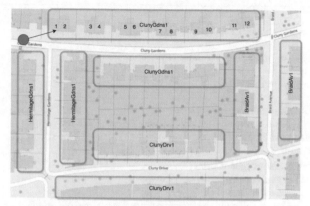

(a) Applying the first pattern based on the first gene *ClunyGdns*1

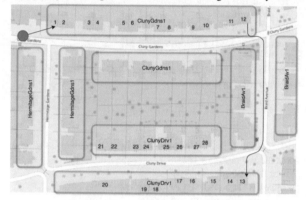

(b) The second gene is a pair *ClunyDrv*1, a pattern covering both sides is applied and added.

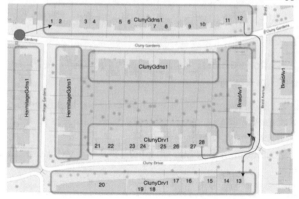

(c) The third gene *BraidAv*1 would now be added

Fig. 11.2 The Application of SBR patterns, based on the genotype shown in Table 11.1. We show the application of the first two genes. The remaining genes are added in the same manner until the entire chromosome has been decoded. *Base map and data from OpenStreetMap and OpenStreetMap Foundation* (https://www.openstreetmap.org/copyright)

Algorithm 22 The SBR decoding algorithm

1: **Procedure** $decoder(genotype, problem)$
2: $deliveries = []$
3: $deliveriesLeft = problem.qtyDeliveries$
4: $geneCount = 0$
5: $prevDelivery = problem.start$
6: $dist = 0$
7: **while** $geneCount < genotype.length$ **do**
8:    **if** $genotype[geneCount] == genotype[geneCount + 1]$ **then**
9:       $sdoubleSided = true$
10:      $geneCount + +$
11:    $current = genotype[geneCount]$
12:    $next = genotype[geneCount \mid 1]$
13:    $lastDel = deliveries.tail()$
14:    $street[]applyPattern(prevDelivery, current, next, doublesided))$
15:    $nextDel = street.head()$
16:    $dist = dist + walkingDist(lastDel, nextDel)$
17:    $deliveries = deliveries.append(street)$
18: $deliveriesLeft = deliveriesLeft - street.size()$
19: $lastDel = deliveries.head()$
20: $dist = dist + (walkingDist(lastDel, problem.end) * delieriesLeft)$
21: $deliveries = deliveries.append(street)$
22: **Return** dist
23: **EndProcedure**

where

- *getWalkingDist(d1,d2)* returns the walking distance between deliveries *d1* and *d2*.
- *<set>.tail()* returns the last item in *<set>*
- *<set>.head()* returns the first item in *<set>*
- *<set>.append(s)* appends the contents of set *s* to *<set>*
- *<set>.size()* returns the number of items in *<set>*

Genetic Operators

The mutation operator employed is relatively simple; it performs one of three randomly selected operations:

- Select a gene at random and move it to another random position.
- Select a gene at random and move it to a position adjacent to another gene which has a junction in common with it.
- Select a gene at random, if it is one of a pair, move it to a position adjacent to its partner gene.

Algorithm 23 Apply SBR Pattern

1: **Procedure** $applyPattern(prevSt, currentSt, nextSt, doublesided)$
2: $startJunction = getNearest(currentSt, prevSt)$
3: $endJunction = getNearest(currentSt, nextSt)$
4: $deliveries = []$
5: **if** $startJunction == endJunction2$ **then**
6:    $sameEnd = true$
7: **if** $!doubleSided$ **then**
8:    **if** $currentSt.side1Done == false$ **then**
9:       $deliveries.append(current.side1Dels$
10:       currentSt.side1Done = true;
11:    **else**
12:       $deliveries.append(current.side1Dels$
13: **if** $doubleSided \&\& sameEnd$ **then**
14:    $deliveries.append(currentSt.side1Dels)$
15:    $deliveries.append(currentSt.side2Dels)$
16: **if** $doubleSided \&\& !sameEnd$ **then**
17:    $deliveries.append(currentStcrossOverDels)$
18: **Return** deliveries
19: **EndProcedure**
where

- *prevSt* is the previous street in the genome (or the start point if *current* is the first street.
- *current* is the street that we are currently routing
- *next* is the following street in the genome or the end point if
- *current* is the last street in the genome
- *<street>.side1Dels* is the set of deliveries on side1 of <street>.
- *<street>.side2Dist* is the set of deliveries on side2 of <street>.
- *<street>.crossOverDels* is the set of deliveries on both sides of <street> crossing as required.
- *<street>.side1done* If the street is being dealt with as two separate sides then side1Done is set to True when the first side is delivered.

The recombination operator selects two parents using a tournament selection of size 2. Let us assume that the two parents have the following chromosomes (where A–H are street sections):

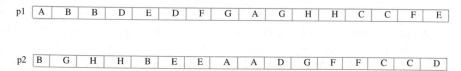

p1 | A | B | B | D | E | D | F | G | A | G | H | H | C | C | F | E

p2 | B | G | H | H | B | E | E | A | A | D | G | F | F | C | C | D

The Child chromosome is created by copying the first gene from *p1* and then each subsequent gene as long as the street sections represented by the gene are connected via a junction directly to the previous gene/street section.

If we assume that A–B, B–D and D–E are linked by common junctions but no direct link exists between D–F, then the following would be copied to the child:

ch | A | B | B | D | E | D | - | - | - | - | - | - | - | - | - | -

The group of genes copied from p1 should form a useful section of route that comprises adjacent street sections. The remainder of the child is created by copying genes from p2. Each gene in p2 is considered and copied to the next free space in the child if it does not compromise the validity of the child. Note that this validity has to take into account that genes can appear once or twice in the chromosome.

This results in the following chromosome:

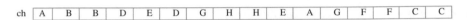

ch | A | B | B | D | E | D | G | H | H | E | A | G | F | F | C | C

The resulting child contains elements inherited from both parents.

Fitness

A significant difference between the original SBR implementation Urquhart et al. (2001) and that presented here is in the fitness function. Within the original, the fitness function was the total distance travelled in order to make the deliveries. We modify that to have the fitness as being the weighted distance travelled *between* street sections. Algorithm 22 line 20 calculates the weighted distance between sections. The fitness function does not take into account the distance covered walking within street sections, but only the distance walking between sections that do not have a common junction. By weighting this distance with the deliveries left, solutions that involve the delivery person carrying large quantities of deliveries between streets are minimised.

This function will tend to favour solutions that deliver outwards from the starting point and then have the delivery person walk back to the starting point after making the deliveries.

11.3 Implementation

A Java implementation of SBR may be found in the repository that accompanies this book. The SBR implementation is based on the classes shown in Fig. 11.3.

The reader should note the relations between *Individual*, *StreetSection* and House, particularly the 3 relations between *StreetSection* and House. Each *StreetSection* has 3 collections of houses as follows:

- *side1*—all of the houses on side1.
- *side2*—all of the houses on side2.
- *bothSides*—all of the houses within the *StreetSection*.

Each of the collections is ordered in walking order from junction 1 to junction 2, the nearest neighbour heuristic being used to determine the order.

We make use of *book.ch7.cache* as the basis for a new *PeristantCache* which creates a local cache of geocoded addresses. We also use classes from the *book.ch7* package within the *RoutingFactory* class which is used to provide *SBRIndividual* with access to mapping data and routes.

Caching Journeys

One aspect of the implementation that is worthy of further discussion is the cache used in *RoutingFactory*. Previous attempts at caching journeys (see Chap. 9) have used an array as the underlying data structure. The array has the advantage that items can be retrieved rapidly from the cache, but it has the disadvantage that the maximum size of the cache needs to be known before starting and that each destination needs to have an index that can be used as the subscript to the array.

When using SBR, we don't need to calculate journeys between every single delivery point (House), but only between street sections. The journey distances between houses within street sections are pre-calculated and stored when the algorithm is launched.

In this implementation, we take a different approach to caching routes. Each route is constructed between two *StreetSection* objects; like all Java objects, each *StreetSection* object has a unique hashCode value allocated to it when it is instantiated. The hashCode of any Java object can be returned as an integer value by calling the *hashCode()* method. Any journey between two street sections can be identified by the pair of hashCodes that represent the StreetSection objects at the start and finish of the journey. We can allocate each journey a unique identifier by combining the hashCodes of the associated street sections using a *Pairing Function* Pigeon, Steven (n.d.). A pairing function accepts two numbers and returns a unique value that combines the two inputs. Using the pairing function allows us to generate a unique id number for each journey, which can then be used as a key to a *HashMap* that contains the cached Journey objects.

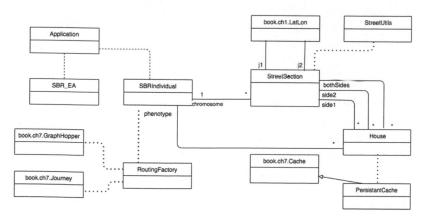

Fig. 11.3 The classes that make up the SBR solver

In our implementation, we use Cantor's Pairing function. Cantor's function takes two integers x, y and returns a third unique integer z. The reader should note two properties of the function:

- For a given combination of x and y, the same value z is always returned.
- Each value of z is only associated with one pair of x and y values.

As z is unique to each pair x and y, it is possible to determine x and y given z, but that does not concern us in this application.

z may be calculated thus:

$$z = (0.5 * (x + y) * (y + y + 1) + y)$$

Thus, for any Journey between two street sections, we can generate a unique journey id based on the Cantor pair of the hashCodes of the two street sections. If the journey id exists within the HashMap, return the *Journey* object, else create a new *Journey* object and add it to the cache as well as return it.

```
1     public double getJourneyDist(LatLon start,LatLon end, String mode)
      {
2         long key = cantorPair(start.hashCode(),end.hashCode());
3         Double res = cache.get(key);
4         if (res == null) {
5             options.put(RoutingEngine.OPTION_MODE, mode);
6             res = router.findRoute(start, end, options).getDistanceKM
      ();
7             cache.put(key, res);
8         }
9         return res;
10    }
11
12    /*
13     * Local cache based on combined hashmaps
14     */
15    private static HashMap<Long,Double> cache = new HashMap<Long,
      Double>();
16
17    private  long cantorPair(long a, long b) {
18        //Use Cantors paring function to generate unique number
19        long result = (long)(0.5 * (a + b) * (a + b + 1) + b);
20        return result; //Return the result
21
22    }
```

Listing 11.1 A cache using Cantors' pairing function to create keys

It should be remembered that retrieval using a *HashMap* object is significantly slower than using an array. But where the size of the cache is likely to be relatively small (e.g. when using SBR), the retrieval speed may be acceptable.

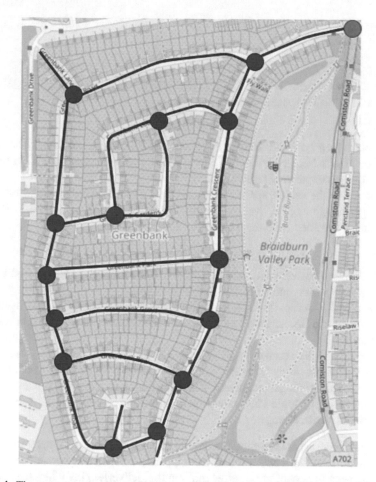

Fig. 11.4 The test area comprising 442 houses within 23 street sections. *Base map and data from OpenStreetMap and OpenStreetMap Foundation* (https://www.openstreetmap.org/copyright)

11.4 Comparison with TSP

As a case study, we examine a collection of streets from the Greenbank area of the City of Edinburgh, UK. The area used for our case study comprises 442 houses within 23 street sections (see Fig. 11.4).

The street sections are set up as per the example in Listing 11.2; for each street section, we specify the house numbers that require a delivery, which are grouped by side. The combination of street name and house number can be used as input to a Geocoder in order to geolocate each house. In a production environment, this data would be held in a file/database and read in at run time.

For an in-depth comparison of TSP versus SBR, the reader is directed to Urquhart, Neil (2002).

```
1        chromo.addStreet(new StreetSection("Greenbank Crescent,
     Edinburgh",new int
     []{39,41,43,45,47,49,51,53,55,57,59,61,63,65,67,69},new int
     []{52,54,56,58,60,62,64,66,68,70,72,74,76,78,80,82}, new
     LatLon(55.9177482, -3.2166029), new LatLon
     (55.915377,-3.216826)));
```

Listing 11.2 Setting up the SBR network in code

TSP

Our problem may be formulated as a 442 city TSP problem which may be solved using the Nearest Neighbour heuristic (Sect. 1.3); the Java implementation may be found in the accompanying software repository.

In this case, the search space of the TSP is $\frac{442!}{2} = 5.487001 * 10^{978}$ based on possible permutations of deliveries. The fitness function used within the TSP is the total distance travelled from the start to the end.

Results

Table 11.3 gives a basic comparison of SBR-EA, TSP-EA and the NN heuristic. Both EAs use an evaluation budget of 100,000 evaluations.

Because SBR-EA uses the distance walked between streets as its fitness, the results presented for SBR-EA are based on the total route distance of the solution with the lowest fitness.

SBR-EA finds the shortest route and does so in the quickest time. The nearest neighbour heuristic finds a longer route using about twice the time. TSP-EA takes considerably longer (approx. 100 times longer) to find a route that is significantly longer. Figure 11.5 shows the best solution found using SBR-EA. Some characteristics worth noting:

- The route encompasses continuous deliveries until the final walk back to the start (shown in red in Fig. 11.5). From the perspective of a delivery person, the longest walking section is carried out after making the deliveries (i.e. when they are not carrying any load).
- The initial street sections (1-4) are dealt with as double-sided which reduces the overall weight carried by making as many deliveries as possible earlier on within the route.

Table 11.3 The route lengths found when using an EA with the SBR representation, an EA with a TSP representation and the nearest neighbour heuristic

Solver	Avg. time (s)	Distance									Min.	Avg.
NN	20.3	10.2	10.2	10.2	10.2	10.2	10.2	10.2	10.2	10.2	10.2	10.2
SBR-EA	9.6	7.3	8.3	8.2	8.1	7.3	7.7	7.3	7.8	8.3	6.8	6.8	7.7
TSP-EA	932.4	21.2	19.2	18.1	22.3	18.4	18.2	22.4	20.3	18.8	21.0	18.1	20

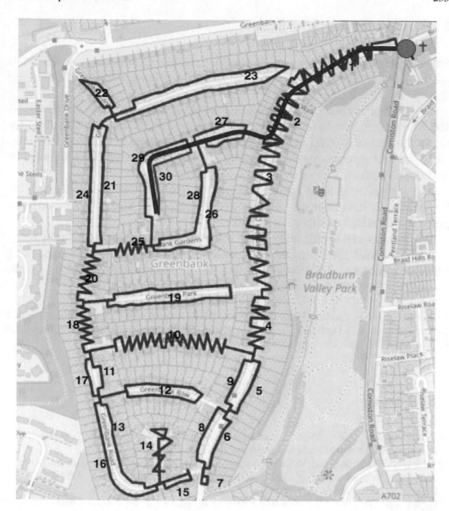

Fig. 11.5 The shortest route found using SBR-EA. The section in red shows street sections that are traversed without making any deliveries (they were delivered earlier). Note that this walk occurs at the end of the route when the delivery person is carrying the least weight. *Base map and data from OpenStreetMap and OpenStreetMap Foundation* (https://www.openstreetmap.org/copyright)

On the basis of this limited study, SBR-EA finds a short route in a reasonably quick time. The nearest neighbour heuristic can find a solution quickly, but, in this case, it is inferior to that found using SBR. Note that if the EA is allowed a sufficiently long run time, then we would expect it to find a solution that equals that found by SBR. The advantage of SBR is the ability to find a solution far quicker than SBR.

11.5   Conclusions

Street Based Routing is a representation and decoder that is tailored towards deliveries within dense urban areas. It exploits the groupings of houses (or other addresses) that occur within streets.

If we accept that delivery addresses that are physically adjacent in location terms should be adjacent within a delivery route, then we can discount any solutions that place them apart within the route. SBR exploits this assumption and keeps deliveries within a street section adjacent, by having no means of representing solutions that would split them up.

It is recognised that the comparison offered within this chapter is limited, but it should show the reader the fundamental advantages of SBR when applied to a problem with a high density of deliveries.

It is recognised that ultimately the TSP-based EA should find the same or better solution than SBR-EA. There may exist some solutions that cannot be created using SBR, but can be with TSP. We assume that the speed advantage of SBR outweighs the possibility of TSP finding a shorter solution.

Acknowledgements The original work on Street Based Routing was undertaken by the author during his time as a Ph.D. student, Urquhart et al. (2001). Thanks are due to Ben Paechter, Peter Ross and Ken Chisholm who supervised the original work.

References

Pigeon, Steven. n.d. Pairing Function, *From MathWorld–A Wolfram Web Resource, created by Eric W. Weisstein.* https://mathworld.wolfram.com/PairingFunction.html
Urquhart, N., B. Paechter, and K. Chisholm. 2001. Street-Based Routing Using an Evolutionary Algorithm. In *Applications of Evolutionary Computing*, ed. E.J.W. Boers, 495–504. Berlin Heidelberg, Berlin, Heidelberg: Springer.
Urquhart, Neil. 2002. *Solving a Real-World Routing Problem using Evolutionary Algorithms and Agents*, PhD Thesis, Edinburgh Napier University, Edinburgh.

Chapter 12
Delivering Parcels

Abstract Reduction of emissions and congestion in city centres is an increasing priority as countries strive to reduce their CO_2 and other emissions. The use of the techniques outlined in earlier chapters can assist in the reduction of emissions for individual problem instances, but the investigation of more radical solutions is desirable in order to achieve greater reductions. The problems described so far within this book are concerned with operational issues, providing solutions for specific problem instances. In this chapter, we review the use of some of the techniques outlined in this book applied to the problem of placing and utilising *micro-depots* (MDs) in a city centre. MDs allow a larger supply vehicle to drop off deliveries at the MD and then low-emission couriers (e.g. walking, cycling or electric vehicles) to be used to make final deliveries. This work is centred around the exploration of a new problem domain rather than solving specific operational instances of a problem.

12.1 Problem Description

Last mile logistics CILT (UK) (2018) have taken on an increasingly high profile as city centres become more congested and polluted. The growth of services such as online shopping has driven increases in deliveries of parcels to home and workplaces, resulting in increased numbers of light goods vehicles (vans) being used within city centres. This has become undesirable for a number of reasons:

- Congestion caused by vehicles driving and parking on city centre streets.
- Pollution caused by vehicles; this is especially acute due to the stop/start nature of this work.
- Increasing costs of deliveries due to delays caused by congestion.

Many cities have introduced low-emission zones or similar measures to reduce congestion and improve air quality; such zones normally prohibit the use of highly polluting vehicles. The use of delivery modes such as walking couriers and cycle couriers has become common along with the use of electric vehicles (EVs). Walking and cycle couriers have a very small environmental impact and contribute little to congestion. As city centres become increasingly pedestrianised, the use of walking

© Springer Nature Switzerland AG 2022
N. Urquhart, *Nature Inspired Optimisation for Delivery Problems*,
Natural Computing Series, https://doi.org/10.1007/978-3-030-98108-2_12

and cycle couriers becomes an attractive option. EVs may contribute to congestion (depending on their size and use) but have no emissions at the point of use.

Because of the limited carrying capacity of walking and cycle couriers, incorporating them into existing logistics networks can be a challenge. One way of utilise such modes is the adoption of *micro-depots* (MDs) Fischer, Michael (2017). A micro-depot is a container where a batch of deliveries may be dropped off by a larger supply vehicle (e.g. a light goods vehicle or van) and stored for transfer to smaller capacity couriers. The MD is sited at a location that allows the supply vehicle to unload and is located close enough to the city centre in order to allow walking and cycle couriers to be utilised for the deliveries. The micro-depot should reduce the number of vans in urban areas leading to reductions in pollution and congestion. Through the adoption of micro-depots, congestion of city streets and CO_2 emissions may be decreased Browne et al. (2011).

We use a fictional scenario based upon the City of Edinburgh; this scenario is similar to that described in Urquhart et al. (2019); the scenario data is available in the repository that accompanies this book. The City of Edinburgh, like many cities, wishes to reduce pollution and congestion in its city centre. One option that may contribute towards that aim is the adoption of micro-depots. Our scenario envisages a courier depot being located to the West of the city from which deliveries are dispatched in the city centre. The scenario that the 10 test problems are based on contains 15 potential micro-depot sites (Fig. 12.1); each of the problem instances contains 50 deliveries randomly selected from a pool of 100 deliveries (Fig. 12.2). Within the set of problems, there is a 50 % chance of a delivery appearing in a particular instance. Under this scheme, although each of the 10 instances is different, they will share some deliveries. This is representative of daily problems, where some customers might receive deliveries on a regular basis. Each solution has 7 characteristics that are of interest to the end user; see Table 12.1.

Table 12.1 The 7 solution characteristics used in this chapter

	Name	Description	Unit
1	Emissions	The emissions generated by the solution	CO_2
2	Time	The time between commencing and making the last delivery	Minutes
3	MDs Used	The number of micro-depots used in the solution	
4	Bikes	The number of cycle couriers used in the solution	
5	Walkers	The number of walking couriers used in the solution	
6	eVans	The number of EV couriers used in the solution	
7	ByMD	The % of deliveries made via a micro-depot	

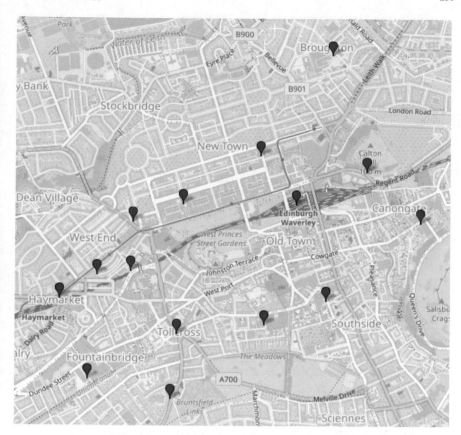

Fig. 12.1 Potential Micro-Depot sites. Base map and data from OpenStreetMap and Open-StreetMap Foundation (https://www.openstreetmap.org/copyright)

12.2 Methodology

Use of MAP-Elites

Within this chapter, we adopt the methodology outlined in Chap. 6 of using an illumination algorithm to create a large archive of possible solutions and then using a parallel coordinates-based tool named ELite VISualisation (ElVis) Urquhart, Neil (2019) to explore the results and to gain some understanding.

From a planning perspective, we may treat the incorporation of micro-depots into a logistics network as an optimisation problem, which seeks to reduce the overall cost or environmental impact. In order to understand how to deploy micro-depots and the associated couriers, it is desirable to understand the range of solutions available and the likelihood of achieving specific targets as well as being aware of the potential for trading off targets such as environmental impact versus cost. In this context, the user is not looking for one specific solution, but a wide range of solutions that collectively

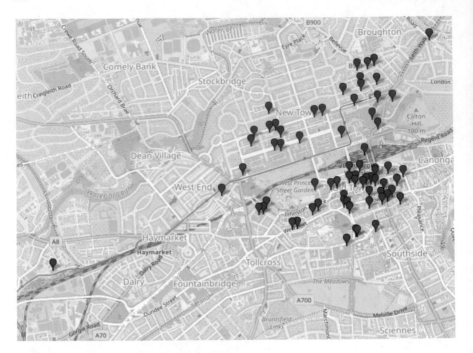

Fig. 12.2 The pool of deliveries used to generate the problem instances. Base map and data from OpenStreetMap and OpenStreetMap Foundation (https://www.openstreetmap.org/copyright)

describe what is possible and what the effects of adopting a particular strategy may be. As described in Chap. 6, Illumination Algorithms are especially well-suited to this type of problem.

Use of Machine Learning to Extract Policies

We can use Parallel Coordinates plots to visualise our archives and highlight solutions that are of interest to us. Another approach is to utilise *Machine Learning* (ML) to automate the extraction of rules from the data within the archives.

ML is normally used to learn a set of rules that allow data examples to be classified. Ideally, these rules will hold on to unseen examples as well. A classical example of ML would be to examine bank loan applications and classify them as those that should be granted (because there is a strong likelihood that the loan will be repaid on time) and those that should be refused. By allowing the ML algorithm to train itself on a set of previous examples where the outcome is known, the algorithm may be able to discover features within the data which will predict the outcome.

We use ML in a slightly different way to that described above; our interest is in finding rules that classify solutions according to criteria (e.g. those solutions with low emissions). We don't have a need to classify unseen solutions; to ease confusion, we refer to rules that the ML produces as *policies*. We define a policy as a set of actions that if followed should lead to a particular outcome, for example, a policy that leads to low emissions might be.

Minimise the use of diesel-powered vehicles AND Maximise the use of bicycles
A policy may be used to guide planners; in this case, the policies are intended to
assist planners in making the best use of micro-depots and couriers.

A wide variety of ML techniques are utilised by practitioners; a comprehensive
discussion is outside the scope of this book—the reader is directed to Forsyth (2019)
for an overview. In this chapter, we use the Flex-GP library Arnaldo et al. (2015),
Group (n.d.). FlexGP has a number of advantages in this context; it is easily available
as a JAR library to download and it makes use of Evolutionary principles and
Pareto dominance; both concepts will be familiar to readers of this book. The author
previously used FlexGP in Urquhart et al. (2021).

Flex-GP initially generates a set of conditions based on each of the decision
variables; the conditions take the format:

$$C_x = 1.0 <= X1 <= 3.6$$

where $X1$ is a decision variable. FlexGP uses *Genetic Programming* to create Boolean
expressions using AND and OR operators, e.g.

$$((C1ANDC2)ORC3)$$

A population of expressions is maintained, and evolutionary operators are used to
modify expressions and create new ones. The fitness of the individual expressions is
based on *accuracy* and *complexity*. Accuracy measures the number of data instances
that the expression correctly classifies; complexity describes the complexity of the
evolved expression. FlexGP evolves a non-dominated front of expressions based
on the trade-off between complexity and accuracy. In this chapter, we adopt the
expression at the "knee" of the front as it will represent the best compromise.

Having generated an archive of solutions using MAP-Elites, the summary (which
contains the list of solution characteristics) is obtained. Assuming that we have 7
problem characteristics, our summary will look something like this:

```
3,10,4,1,1,2,3
1,4,2,10,6,8,7
8,6,5,5,6,10,2
2,7,1,10,3,2,1
...
```

Let us assume that the first characteristic is emissions and that we wish to classify
solutions as very low emissions or not. Assuming that we define low emissions as a
score of 1 or 2, we can then add a classification to each solution (1== low emissions,
0== not low emissions). Removing the emissions column (otherwise, the GP will
just find a rule that states low emissions are based on low emissions ...)

```
10,4,1,1,2,3,0
```

Table 12.2 The labels used when translating policies into text

Range	Description
1–2	Very Low
3–4	Low
5–6	Medium
7–8	High
9–10	Very High

```
4,2,10,6,8,7,1
6,5,5,6,10,2,0
7,1,10,3,2,1,1
...
```

where
1 = very low emissions.
0 = higher emissions.

The task of FlexGP is now to find out what data patterns in cols 1–6 predict the classification (0 or 1) in col 7. When translating the policies into human-readable policies, it is useful to have labels that can be attached to the problem characteristics values (see Table 12.2).

12.3   Implementation

The MAP-Elites algorithm (see Chap. 5) used to produce the results in this chapter is a development of that originally outlined in Urquhart et al. (2019). The ElVis tool Urquhart, Neil (2019) is available for use online.[1]

Our implementation is based around the concepts of an *MDProblem* class which encapsulates problem details, a *Model* class that is used to convert a genome (see below) into a solution; the *Model* class is used to simulate the logistics network and produce a *Results* object which contains the solution (see Table 12.1). Within the MAP-Elites algorithm, the archive is a 7-dimensional space, based upon the solution characteristics. Each characteristic is normalised on a scale of 1–10, which gives a maximum archive size of 10^7.

Representation

Our representation is based upon the concept of commencing with a default delivery plan based upon no micro-depots being used (the *default tour*). The default tour represents a viable, although possibly undesirable, solution that does not make use

---

[1] https://www.commute.napier.ac.uk/upload.

of any micro-depots or couriers. We seek to find a set of changes to the default tour which when applied will improve the tour through the use of micro-depots and couriers. As the default tour undertakes all of the deliveries using one vehicle, it can be considered an instance of the TSP. The nearest neighbour heuristic is used to construct the default tour during the initialisation phase of the algorithm.

Each solution is constructed by starting with the default tour and applying a series of changes that are specified by the genes in the chromosome. Each gene represents the addition of one courier which will operate from a micro-depot making 1 or more deliveries.

Each gene contains the following information:

- **Courier type** (bike | walk | electric van).
- **Delivery Count** the number of deliveries to be transferred to the courier.
- **Micro-depot** the id of the micro-depot that the vehicle will operate from.

Consider the following gene:

Courier	Depot	Qty
walking	MD2	3

This would be interpreted as follows: create a walking courier based at micro-depot 2, which makes 3 deliveries.

Evolutionary Operators

When creating a new chromosome during the initialisation phase, a random number (0–10) of genes are created with randomly set values.

Chromosomes are mutated by selecting one of the following actions at random:

- Add a new gene to the chromosome (based on random values).
- Randomly alter the chromosome order, by randomly selecting and moving a gene.
- Randomly select and mutate an individual gene (by altering one of its four values).
- Randomly delete a gene from the chromosome.

A new chromosome is created by copying 10 genes each of which is selected from a random position within a randomly selected parent.

Decoding and Evaluation

A possible chromosome with 4 genes might look like this:

Courier	Depot	Qty
walking	2	3
cycling	1	4
EV	3	4
walking	1	3

Algorithm 24 Create courier tour

1: **Procedure** $createCourierTour(Tour supplyTour, Courier c Micro Depot md, int len)$
2: $tour = newTour()$
3: $tour.courier = c$
4: $tour.start = md$
5: $Location current = tour.start$
6: $count = 0$
7: **while** $count < len$ **do**
8:    $Location del = findClosest(supplyTour, current, c)$
9:    **if** $tour.canAdd(del)$ **then**
10:       $tour.add(del)$
11:       $supplyTour.remove(del)$
12:    $count + +$
13: $Location Insert P = f in closest(supplyTour, md, supplyTour.vehicle)$
14: $supply.insertAfter(insert P, tour)$
15: **EndProcedure**

where

- *findClosest(tour,loc,v)* searches through *tour* and returns the location within the tour closest to *loc*. Distances between locations are based on using a vehicle of type *v* (cycle and walking couriers can take advantage of paths not available to larger vehicles).

Each set of courier deliveries is constructed by using the nearest neighbour heuristic as shown in Algorithm 24.

Let us assume that we have 20 deliveries; the solution is decoded from the chromosome as follows:

1. We begin by constructing the default tour using the nearest neighbour heuristic into a default tour as follows (for clarity, we show the NN order as 1–20):

Depot	Vehicle/Courier	Deliveries
D0	Supply	1 2 3 4 5 6 7 8 9 10 11 12 13 14 15 16 17 18 19 20

We assume that the depot used by the default tour is *D0* and that the vehicle used for deliveries on the default tour is the *Supply Vehicle*. In this default solution, the supply vehicle makes all of the deliveries directly. Although this is a valid solution, it may not be a desirable solution.

2. We decode the first gene in the chromosome.

Courier	Depot	Insert	Qty
walking	**2**	**4**	**3**
cycling	1	2	4
EV	3	12	4
walking	1	16	3

The first gene creates a walking courier route from MD2. Customers from the default route are transferred, starting at position 4. Customers are transferred to the courier until either the specified quantity (3) has been transferred or the capacity of the courier is reached. Finally, the default route is modified to include the call to MD2.

Depot	Vehicle/Courier	Deliveries
D0	Supply	1 2 3 MD2 7 8 9 10 11 12 13 14 15 16 17 18 19 20
MD2	Walking	4 5 6

3. Now we decode the second gene:

Courier	Depot	Insert	Qty
walking	2	4	3
cycling	**1**	**2**	**4**
EV	3	12	4
walking	1	16	3

The second gene adds a cycle courier operating from micro-depot MD1.

Depot	Vehicle/Courier	Deliveries
D0	Supply	1 MD1 MD2 8 9 10 11 12 13 14 15 16 17 18 19 20
MD2	Walking	4 5 6
MD1	Cycling	2 3 7

4. The 4th and 5th genes add two additional couriers.

Courier	Depot	Insert	Qty
walking	2	4	3
cycling	1	2	4
EV	**3**	**12**	**4**
walking	1	16	3

Note that only one visit is required to MD1 to deliver the items for both of the couriers that will operate from there.

Depot	Vehicle/Courier	Deliveries
D0	Supply	1 MD1 MD2 8 9 11 14 15 MD4 20
MD2	Walking	4 5 6
MD1	Cycling	2 3 7
MD4	EV	16 17 18 19
MD1	Walking	10 12 13

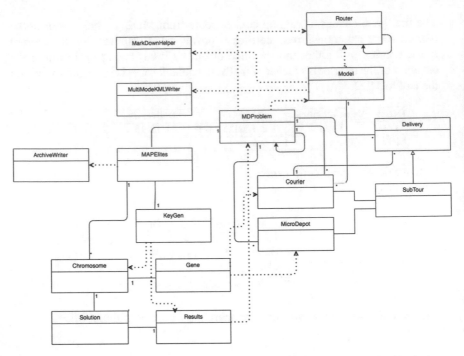

Fig. 12.3 The structure of the Java implementation of the MDVRP solver, used in this chapter

Java Implementation

The Java implementation may be found in the code repository that accompanies this book. Figure 12.3 shows the structure of the solver. The implementation of the algorithm within the *MapElites* and *KeyGen* classes are based on the implementation presented in Chap. 6.

The *MDProblem* class is used to construct a solution from a *Chromosome* object and return the results in a *Results* object.

12.4 Results

To visualise the results, we use the ElVis tool Urquhart, Neil (2019), which produces a parallel coordinates chart and a heatmap for each pair of solution characteristics.

As we are interested in finding out more about how MDs could be used, we aggregate the results for the 10 problem instances. Figure 12.4 shows the parallel coordinates plot of the aggregated archives and Fig. 12.5 shows heatmaps for each pair of coordinates.

The heatmaps give us an indication as to which areas of the solution space are not filled. It would be naive to suggest there are no viable solutions in these gaps,

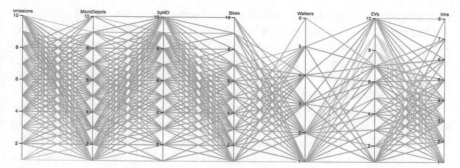

Fig. 12.4 A parallel coordinates plot based on the combined archive

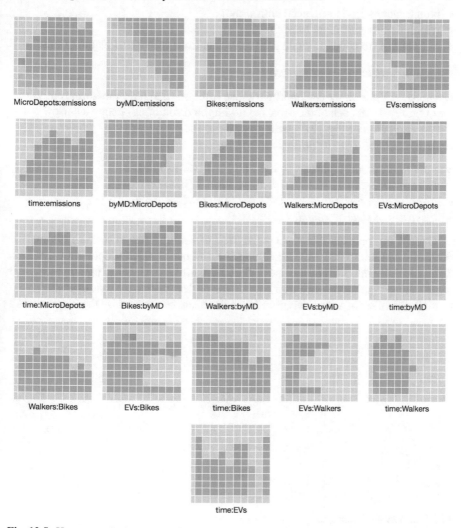

Fig. 12.5 Heatmaps plotting each pair of solution characteristics

but the algorithm used is unable to find any in the run time allowed. One of the most obvious gaps in the solution space is that shown by the relationship between the use of micro-depots (shown as byMD—the % of deliveries made via a micro-depot) and emissions; the heatmap suggests that there are no solutions with low emissions and low micro-depot use. The triangular nature of the heatmap suggests a very specific relationship between the two characteristics.

Our interest lies not in finding a solution to one or more specific problem instances, but rather it lies at the policy level, finding out more about the effects of incorporating MDs and associated couriers into an existing logistics network. We aim to use the archive of solutions to find answers to the sort of questions that may be posed by policy makers:

- Which factors lead to reduced emissions?
- Which factors lead to reduced time?
- Is it possible to combine rapid deliveries (i.e. low time) with low emissions?

Which factors lead to reduced emissions?

Using the ElVis tool, we can highlight those solutions that have the lowest emissions (see Fig. 12.6). On each of the axes, the highlighted solutions occupy a narrow band. An interpretation of Fig. 12.6 could give rise to the following:
Emissions are likely to be lower in solutions that

- *send the maximum quantity of deliveries via MDs (byMD axis).*
- *use the minimum number of walking couriers (Walkers axis).*
- *use a smaller number of Micro-Depots (Micro-Depot axis).*
- *make use of electric vehicles (EVs axis).*

One of the policies that emerges is that of sending the maximum number of deliveries via micro-depots, but keeping the number of micro-depots in use small. This would be supported by the use of Electric Vans which have a larger payload and higher speed than the other courier types. Thus, a fleet of electric vans operating from a smaller number of micro-depots reduces the distance covered by the larger supply

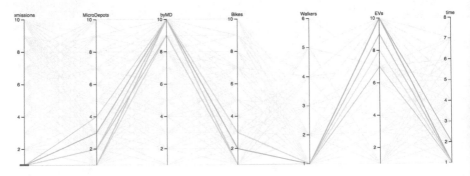

Fig. 12.6 Low-emission solutions

vehicle when delivering to the micro-depots and maximises the distance covered by the EVs.

Let us now examine the policy determined by GP. As discussed earlier, Flex-GP creates a Pareto front of policies that trade-off accuracy versus rule complexity. For each of our policies, we will examine the "knee" solution which should offer the best compromise.

The Raw policy is

$$(1.0 <= X1 <= 3.6 \text{ and } 7.5 <= X2 <= 10.0)$$

where
$X1$ is Micro Depots.
$X2$ is % by MD.

By substituting variable names and labels (from Table 12.2), our policy becomes

$$(MicroDepts <= Low) \text{ and } (ByMD >= High)$$

The literal interpretation of this policy could be
Send the maximum amount of deliveries via the minimum number of micro-depots.
In practice, this policy makes the maximum use of low-emission couriers and reduces the length of the tour that the supply vehicle has to undertake.

Which factors lead to reduced delivery time?

Figure 12.7 highlights low time solutions; unlike emissions there is not a clear pattern of solutions with specific characteristics that lead to a low delivery time. What can be discerned from Fig. 12.7 is that lower numbers of walking couriers are more likely to result in faster delivery times—this would make sense given that walking couriers are the slowest form of delivery open to us.

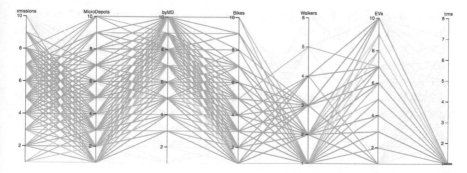

Fig. 12.7 Low delivery time solutions

If we run Flex-GP on the same problem, we can extract the raw policy (based on the knee solution):

$$((1.0 <= X5 <= 8.6 \text{ and } 1.0 <= X4 <= 2.0)$$

$$and$$

$$(1.0 <= X2 <= 8.4 \text{ or } 1.0 <= X3 <= 2.7))$$

where
$X2$ is % by MD.
$X3$ is Bikes used.
$X4$ is Walkers used.
$X5$ is EVs used.
By substituting variable names and labels (from Table 12.2), our policy becomes

$$((EVsUsed <= VeryHigh \text{ and } WalkersUsed <= VeryLow)$$

$$and$$

$$(\%ByMD <= High \text{ or } BikesUsed <= Low))$$

The literal interpretation of this policy could be
Make maximum use of EVs and make minimum use of walking couriers. And send as many deliveries via MDs or minimise the use of bicycles.
Overall, the policy suggests maximum use of MDs and minimum use of walking couriers and bicycle couriers. This makes some sense; by maximising the use of micro-depots, we have more goods being delivered by couriers who work concurrently. The fastest couriers are those using EVs so they are prioritised over Walkers and Cyclists who are considerably slower.

Is it possible to combine rapid deliveries (i.e. low time) with low emissions?

Figure 12.8 shows the effect of highlighting those solutions with low emissions and a faster delivery time. We note that Fig. 12.8 is identical to Fig. 12.6. This comes about because the set of solutions shown in Fig. 12.8 is a subset of those shown in Fig. 12.7. In this particular example, it could be said that the emissions solution characteristic overrides time.

Using Flex-GP to analyse the data presented in Fig. 12.8, the following policy is discovered:

$$1.0 <= X1 <= 3$$

where
$X1$ is Micro-Depots Used.

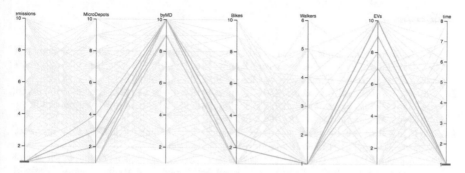

Fig. 12.8 Solutions with low emissions and faster delivery times

By substituting variable names and labels (from Table 12.2), our policy becomes

$$(MicroDeptsUsed <= Low)$$

The literal interpretation of this policy is to reduce the number of micro-depots.

12.5 Conclusions

In this chapter, we considered the Micro-Depot routing problem, which is based upon Urquhart et al. (2019). Our interest is not in whether we can solve specific instances of the problem, but in finding out what the effects of adding micro-depots and the associated couriers might be.

We use the Map-Elites illumination algorithm Mouret and Clune (2015) to generate archives of solutions. We conduct our experiments on a range of problem instances that are each a subset of a larger set of deliveries; each instance could represent a day's deliveries.

Besides using Parallel Coordinates to explore the solutions returned by MAP-Elites we use machine learning to discover policies that help explain the relationships between the solution characteristics.

Acknowledgements This chapter is based upon work originally published in Urquhart et al. (2019, 2021). The author wishes to thank Silke Höhl, Emma Hart and Achille Fonzone for their contribution to the original papers. Thanks are also due to William Hutcheson who undertook much of the development work on the ElVis tool.

References

Arnaldo, I., K. Veeramachaneni, A. Song, and U.-M. O'Reilly. 2015. Bring Your Own Learner: A Cloud-Based, Data-Parallel Commons for Machine Learning. *IEEE Computational Intelligence Magazine* 10 (1): 20–32. http://ieeexplore.ieee.org/document/7010434/.

Browne, M., J. Allen, and J. Leonardi. 2011. Evaluating the use of an urban consolidation centre and electric vehicles in central London. *IATSS Research* 35 (1): 1–6. https://www.sciencedirect.com/science/article/pii/S038611121100015X.

CILT (UK). 2018. The future of last mile logistics. https://ciltuk.org.uk/News/Latest-News/ArtMID/6887/ArticleID/18808/The-future-of-last-mile-logistics.

Fischer, Michael. 2017. *City Logistics: Determination of Criteria to Evaluate Locations for Micro-Depots in Frankfurt am Main*. Master's thesis, Frankfurt University of Applied Science, Frankfurt am Main.

Forsyth, D. 2019. *Applied Machine Learning*, Springer International Publishing, Cham. http://link.springer.com/10.1007/978-3-030-18114-7.

Group, A. n.d. Rule Tree learner. http://flexgp.github.io/gp-learners/ruletree.html.

Mouret, J.-B. and Clune, J. 2015, Illuminating search spaces by mapping elites, (1504.04909). _eprint: 1504.04909. https://arxiv.org/abs/1504.04909.

Urquhart, N., Höhl, S. and Hart, E. 2019. An illumination algorithm approach to solving the micro-depot routing problem, *in* A. Auger and T. Stützle (eds), *Proceedings of the Genetic and Evolutionary Computation Conference, GECCO 2019, Prague, Czech Republic, July 13-17, 2019*, ACM, pp. 1347–1355. https://doi.org/10.1145/3321707.3321767.

Urquhart, N., Höhl, S. and Hart, E.: 2021, Automated, Explainable Rule Extraction from MAP-Elites Archives, *in* P. A. Castillo and J. L. J. Laredo (eds), *Applications of Evolutionary Computation - 24th International Conference, EvoApplications 2021, Held as Part of EvoStar 2021, Virtual Event, April 7-9, 2021, Proceedings*, Vol. 12694 of *Lecture Notes in Computer Science*, Springer, pp. 258–272. https://doi.org/10.1007/978-3-030-72699-7_17

Urquhart, Neil: 2019, ELite VISualisation (ELVIS). https://commute.napier.ac.uk/upload

Chapter 13
Conclusions and Future Developments

Abstract This book discusses delivery problems and possible means to solve them. The need to reduce carbon emissions coupled with an increasing demand for products and services to be delivered to the home will increasingly require solutions that are further optimised to make better use of low-emission travel modes and allow cooperation between organisations with delivery problems. This chapter carries out a brief review of potential future developments in terms of developments in computing science, transportation and the expectations of consumers.

13.1 Software, Algorithms and AI

All of the software systems described in this book are written to solve specific problems within the domain of delivery problems. In order to produce a solver, it is first necessary to understand the problem and possess the knowledge to create and configure a solver. In a domain where almost every problem may have a slightly different formulation, due to differences in organisations and working practices, we find ourselves in the situation where every problem variant seems to require its own solver. Approaches to coping with this could include

- **Develop a custom solver for each variant**
  This might be viewed by many as the ideal means of solving the problem, but the financial costs of such a development project may be beyond the means of smaller organisations.
- **Develop a complex solver with sufficient options as to allow for the solving of many variants**
  It is technically feasible to design and implement a solver that can solve a range of different problem variants, but the resulting software may be complex to use. The solver software should be usable by domain experts (i.e. those responsible for planning deliveries). Not only does the solver become complex to use, but the source code and design may also become bloated. Despite adding many options to the software, there is still the risk that it will not solve the exact problem that the user needs to solve.

N. Urquhart, *Nature Inspired Optimisation for Delivery Problems*,
Natural Computing Series, https://doi.org/10.1007/978-3-030-98108-2_13

- **Adapt the problem to fit an existing solver**

  For some organisations, this might be the most realistic option, adapt working practices to suit an existing solver that is available to the organisation. But this is an unsatisfactory outcome in many cases and could end up stifling local initiative organisations and practices.

None of the above approaches is an ideal answer to the problem. One possible solution is that of *Artificial General Intelligence* (AGI) (Kurzweil 2005). AGI proposes the construction of an intelligent problem solver that does not need to have domain concepts designed into it, but has the capability of learning about a problem domain in the manner of a human and adapting to solve specific problems. Such a system would be capable of having the problem explained to it and then learning the mapping between the problem and the desired solution.

It may be argued that, at the time of writing, we are still some way off from AGI, but a compromise might exist in the form of *domain-specific AI* (Johar 2019). Such a solver would understand the underlying principles of a domain, but be able to learn from users, by example, how to understand a problem and what characteristics define an acceptable solution. For example, in our field of deliveries most problems are underpinned by a form of TSP—they require the ability to place a set of deliveries into an efficient order. Other concepts such as pathfinding are also common across many problems. It becomes feasible to build a solver that could contain a taxonomy of delivery problems; as a particular problem is explained, the taxonomy recognises and categorises characteristics. As concepts and problems are recognised, appropriate algorithms can be added from libraries. By adopting an illumination approach, the system seeks to construct an archive of solutions. By having the user describe solutions that are of interest to them, key solution characteristics can be identified and used to highlight solutions of potential interest to the user. Ultimately machine learning could be employed to learn from the user what solution characteristics are desirable, then use that to identify solutions from within the archive which will be of interest to the user.

13.2   Developments in Transportation

Recent trends have seen a move towards the use of active travel modes for deliveries, in particular bicycle couriers and walking couriers. Electric vehicles have also had an impact. In many respects, these modes only differ from traditional vehicle deliveries in terms of characteristics such as speed, range and payload. The use of drone technologies for deliveries (Koshta et al. 2021; Moshref-Javadi and Winkenbach 2021) has the potential to disrupt existing logistics models. Drones are not restricted by existing road networks and associated infrastructure, and they are by their nature largely autonomous. The challenges in the use of drones lie not only in safely controlling the flight of individual drones, but also in safely routing numbers of them over cities. Without the structure of a road network to manage them, drones will require means to avoid collisions and safely coordinate their actions. One possibility is the use of techniques from *Swarm Intelligence* (Akram et al. 2017; Spanaki et al. 2021) to manage and coordinate drone deliveries. The problem becomes more complex when one considers that drones may be operated by a number of independent, competing organisations, despite which coordination of actions is required in order to avoid collisions.

A growing trend in recent years has been the use of public transport networks to carry goods, for instance, the carriage of parcels or groceries on a bus or tram alongside regular passengers (Pietrzak and Pietrzak 2021; De Langhe et al. 2019). A major advantage of these schemes is their ability to make use of surplus capacity. A disadvantage is that such links can only form part of a logistics network given that public transport tends to operate to a fixed schedule along predefined routes. There is also the challenge of being able to load and unload goods without causing undue delays and disruption within the public transport network. Possibly, the use of micro-depots in this context, with public transport supplying the MDs might be a fruitful area for future study.

13.3 Consumers

In many respects, the demand from consumers is likely to be the greatest influence on future developments. Increasing demand for home deliveries especially during the COVID-19 lockdown has encouraged many smaller businesses to offer home deliveries as an alternative to face-to-face shopping. This growth of online purchasing has occurred at a time when public awareness and concern over climate change is increasing. Thus, the optimisation of logistics has never been so vital given the challenge of increasing deliveries and reducing environmental impact.

An area that will require increasing investigation is that of cooperation between those undertaking deliveries. Although individual organisations may be in competition when selling goods, it may be in the best interests of society if they cooperate over the deliveries of goods. By sharing vehicles and resources (such as micro-depots), a global benefit may be accrued.

References

Akram, R. N., Markantonakis, K., Mayes, K., Habachi, O., Sauveron, D., Steyven, A., and Chaumette, S. 2017. Security, privacy and safety evaluation of dynamic and static fleets of drones. *2017 IEEE/AIAA 36th Digital Avionics Systems Conference (DASC)*, pp. 1–12. ISSN: 2155-7209.

De Langhe, K., Meersman, H., Sys, C., Van de Voorde, E., and Vanelslander, T. 2019. How to make urban freight transport by tram successful?. *Journal of Shipping and Trade* **4**(1), 13. https://doi.org/10.1186/s41072-019-0055-4

Johar, P. 2019. Council post: How AI can reach its potential by being domain-specific. https://www.forbes.com/sites/forbestechcouncil/2019/08/16/how-ai-can-reach-its-potential-by-being-domain-specific/

Koshta, N., Devi, Y., and Patra, S. 2021. Aerial bots in the supply chain: A new ally to combat COVID-19. *Technology in Society* **66**, 101646. https://www.sciencedirect.com/science/article/pii/S0160791X21001214

Kurzweil, R. 2005. *The singularity is near: When humans transcend biology*. New York: Viking.

Moshref-Javadi, M., and Winkenbach, M. 2021. Applications and research avenues for drone-based models in logistics: A classification and review. *Expert Systems with Applications* **177**, 114854. https://www.sciencedirect.com/science/article/pii/S0957417421002955

Pietrzak, O., and Pietrzak, K. 2021. Cargo tram in freight handling in urban areas in Poland. *Sustainable Cities and Society* **70**, 102902. https://www.sciencedirect.com/science/article/pii/S2210670721001906

Spanaki, K., Karafili, E., Sivarajah, U., Despoudi, S., and Irani, Z. 2021. Artificial intelligence and food security: Swarm intelligence of AgriTech drones for smart AgriFood operations. *Production Planning & Control* 1–19. https://doi.org/10.1080/09537287.2021.1882688

Index

Printed in the United States
by Baker & Taylor Publisher Services